Not all ringers and cowboys

For Sally
My full-time travelling partner

and

In memory of Pete
A fellow traveller, friend and inspiration who seized every moment and lived life
as the adventure it is.

Not all ringers and cowboys

DREW RADFORD

ABC Books

Published by ABC Books for the
AUSTRALIAN BROADCASTING CORPORATION
GPO Box 9994 Sydney NSW 2001

Copyright © Drew Radford 2005

First published February 2005

All rights reserved. No part of this publication may be reproduced, stored in a retrieval system or transmitted in any form or by any means, electronic, mechanical, photocopying, recording or otherwise, without the prior written permission of the Australian Broadcasting Corporation.

National Library of Australia
Cataloguing-in-Publication entry
Radford, Drew.
Not all ringers and cowboys.

ISBN 0 7333 1270 5.

1. Radford, Drew-Journeys. 2. Motorcycling. 3. Australia-Description and travel. I.Australian Broadcasting Corporation. II. Title.

919.4

Photographs by Drew Radford
Designed by i2i design
Cover design by i2i design
Colour reproduction by PageSet, Victoria
Printed by Tien Wah Press, Singapore

5 4 3 2 1

Thanks

The way I usually get things done is by biting off the biggest possible chunk and then frantically masticating until I choke. This approach however wasn't possible with the Bloke on the Bike: it was too big a dish and needed to be tackled by many people chewing in unison. Without writing another book, I can't adequately detail how important all the following people were, nonetheless I wish to take the opportunity to thank…

Nicki for motivating me, over a bowl of pasta, to get back on the road and chase my dreams. My then boss Geoff who not only rejected my resignation but also got behind the idea, built it into something bigger and made it happen. Ian at ABC New Media and Sue at ABC Radio who were both brave enough to support the concept and reach for their cheque books—and Lucy for making sure the cheques were in the post. Special thanks also to DETE SA for supporting both journeys, particularly Kevin, John and Yvonne who made everything happen. The team at Mission Central, a.k.a. ABC Education, who helped make the Bloke on the Bike a reality—particularly Sonja and Lynn who provided support while I was on the road and also Team Carols and Wolf who made sense of the on-line material I piped back. Victor and Angelo for technical support and advice, and also for Victor's audio engineering wizardry in putting the radio series together. Justin at BMW who was brave enough to lease us two bikes in a row (which was remarkable considering the evil things he had witnessed me do to my own bike during our ride back from England). Bruce for all his help in sorting out the first bike and more importantly for his input in getting my riding up to speck—a couple of thousand kilometres of swallowing Bruce's dust and watching how he rides probably saved my butt countless times over. Captain Chris for sorting out everything with Bike 2 and all the extra miles he and the crew at the now defunct Bike City put into keeping me on the road—particularly Dan, Truck, Gustav, Aaron and Jan. Pip for her long suffering support while I was on the road. John for all the help rigging up Bike2. The team at ABC Enterprises, particularly Jill, Geraldine, Helen and Mignon for their guidance. Bernadette, also at Enterprises, for getting the ball rolling by commissioning the audio series. My family, because without the right building blocks (the support and encouragement they have given me since year dot), I would never have been able to take on any of these wacky ventures. And thanks to the people who followed the journeys and sent me emails or came up and said g'day or left notes on my bike, after all you are who it was all for. And finally, and most importantly, to my Sally for her unerring love, encouragement and belief while I slaved over the keyboard…And Jet, yes I am now going to come out of the study and kick the ball.

Contents

	The bit before the start	1
1	Super glue miracles	3
2	How did it come to this?	8
3	Not made in Japan	11
4	It's got nothing to do with pasta	24
5	Mincemeat kneecaps	36
6	Bouncing bombs	42
7	Permit roulette	47
8	Photo failures	57
9	Semi-retirement and cross-dressing	67
10	Urban living in the dead heart	75
11	Passing aliens	82
12	Dressed to build	89
13	Glass art dreaming	94
14	Navigation by braille	101
15	Tar school	116
16	The shire with no town	127
17	Outback title fights	137
18	'I wouldn't trust the maps out here, mate'	146
19	Thirty three thousand litres of diesel	157
20	Twenty-first century swagman	171
21	Petrol pump policies	175
22	Broome-time	183
23	Team Dutch Commodore	189
24	Dirt surfing	198
25	Deserts one month, tropics the next	216
26	Satellite classrooms	230

27	Refrigerated fences	237
28	Five of the seven most deadly	243
29	The Commodore of Birdsville	249

Sprechen sie what?	258
Tech stuff	259
Trainspotting stats	265
A day in the office	272
More trainspotting stuff	278

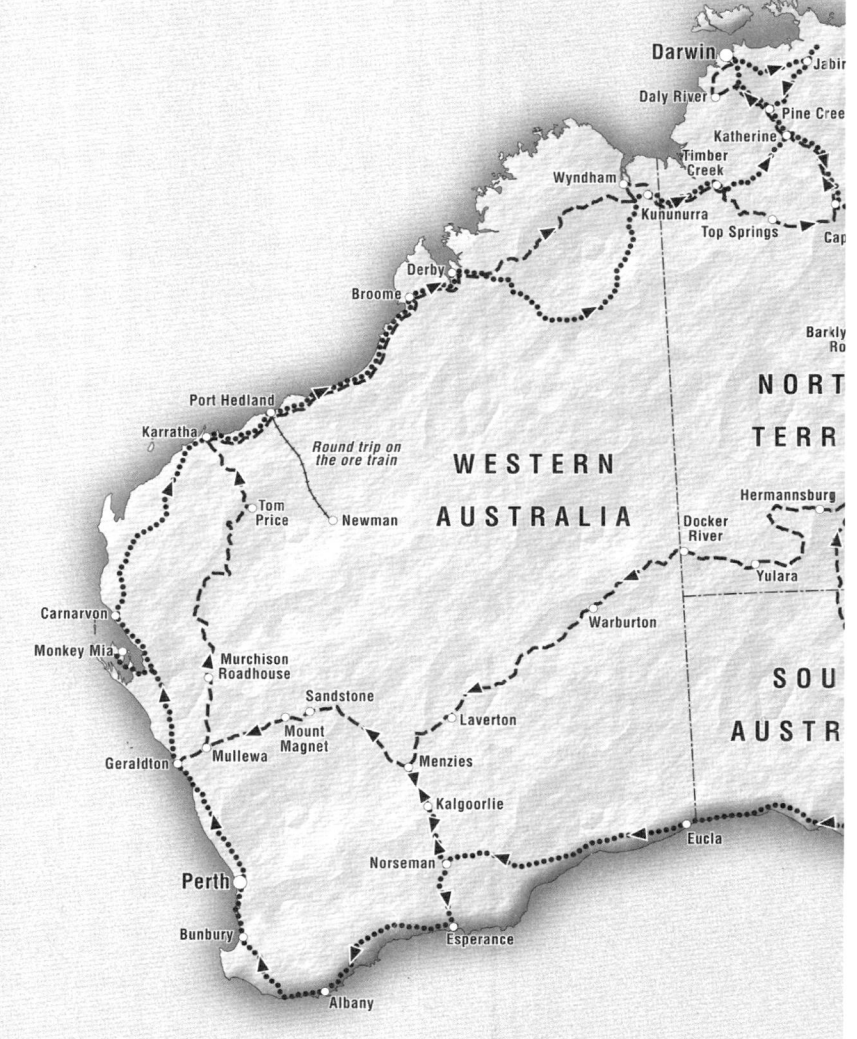

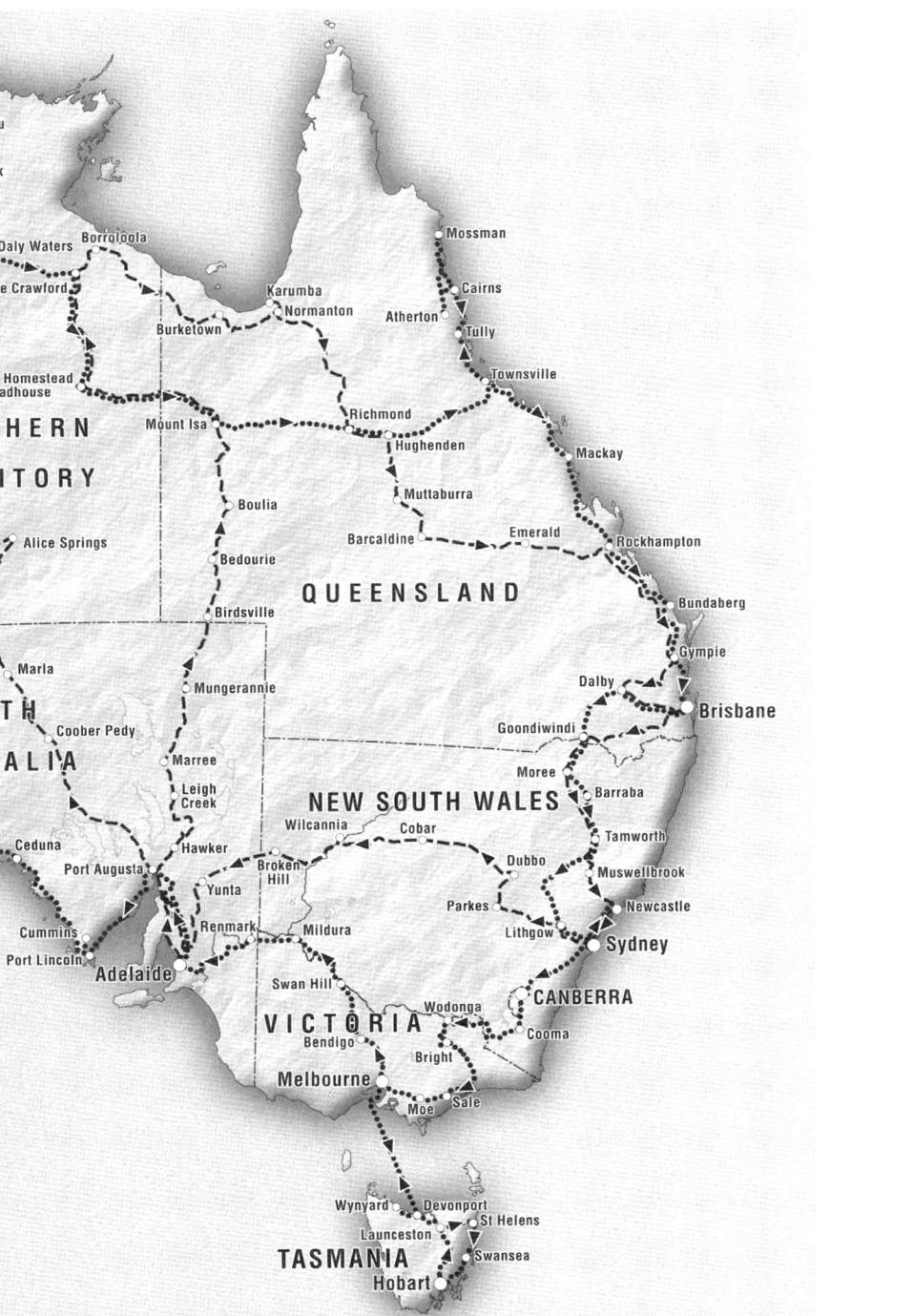

The bit before the start

If you are cursed with a numeric fetish, and if digits ring your bell, then brace yourself. This book is the result of two motorcycles, 11 tyres, 3266 litres of fuel and 51 371 kilometres travelled on two laps of Australia over two years to gather stories for the ABC.

But if you're into something more satisfying than trainspotting-type details like this, then you will be pleased to know that this book has little to do with numbers but lots to do with people's stories, yarns I came across while bashing around the bush in my second year of travel. This second lap concentrated on unearthing people and places that were contrary to the stereotypes we urban-dwellers are sold about the outback; that is, it's not all flaming ringers and cowboys. Marketing people continually sell us this type of imagery—and it's a load of bollocks. These marketers are the same people who recently re-branded Northern Australia's Wet Season as the 'Green Season', justifying the change by saying, 'We feel the term "Wet Season" has negative connotations and it puts tourists off. "Green Season" is more representative because everything is so green at that time of the year.' Damn right, it's green. That's because it's so flaming wet! These are the same clowns that fill every outback tourist brochure with Akubra-wearing stockmen eating 12-kilo steaks off barbies made out of old 44-gallon drums. It all goes to prove one of the world's greatest furphies: 'Hi! I'm from marketing and I'm here to help.'

Having said all that, I am not dopey enough to deny that stockmen and pastoralists were fundamental in opening up a lot of the remote country for white fellas. However, things change and there is now a far broader mix of people living and working in the outback. Nowhere was this clearer than when I was visiting Longreach, home to the Stockman's Hall of Fame. I had been asked to the local school to give a bit of a chat—and out of a class of about 20, only two of the students came from families that worked on the land. And this was in a town that was supposedly one of the great outback meccas. Sure, Longreach is a service town, but the show of two hands definitely meant that ringers and cowboys were in the minority.

This book is primarily about some of the other people who live and work in the outback. It's also about the other travellers you meet as you rattle around the whopping great chunk of dirt that is Australia. Finally, it is all held together by the misadventures of an ill-prepared bloke who probably shouldn't have been out there all alone—but he's damn glad he was.

L to R: Stockman Hall of Fame—Longreach
Central Queenslanders taking the fun out of getting lost
Another flamin' Black Stump;
Cockburn Range —Gibb River Road, W.A.

POSTSCRIPT The slang I use caused a bit of confusion among those who helped put this book together. If you come across words like 'hardcore' and think I am talking about John Holmes and his porno mates, I suggest you check out the list of colloquialisms and definitions on page 258. Also in the back is a bunch of stats and info about the journey and the bike.

Super glue miracles

In the outback, rear-view mirrors are generally only good for double-checking what you've just run over. I was putting mine to a new-found use as I stared at 15 grand's worth of equipment bouncing along the track behind me. 'Bounce' is a little generous; 'cartwheel' may be more to the point. Come to think of it 'stare' was probably also a little off the mark. When you're doing the metric ton, perched on the back of a few hundred kilos of bucking bike and luggage (admittedly it was a little less now that some of the luggage had broken off and was making its own way), you don't have a lot of time to 'stare' in the mirrors. If you do, you end up resurfacing the road with bits of flesh. I guess 'glimpse' would be closer to the truth, just a few sickening ones of my top luggage box looping end over end across the desert, trying to make up for lost ground.

The separate travel itinerary for my luggage was my own fault. I should have taken precautions 50 k's earlier when the track became a lot rougher—violently rough. Jagged outcrops of rocks were ricocheting the bike from right to left, left to right, occasionally bucking it and sending the rear wheel spinning wildly through the air. When the road tired of treating me like a pinball, bulldust-camouflaged ruts took over and launched surprise attacks on my spine. Vertebrae crunched and compressed at each hole I smashed through. I was convinced I would be a few centimetres shorter by the end of the day.

To be honest, I was a little surprised by the state of the track, especially as I didn't really consider it to be all that remote. After all, I was just taking a bit of a dirt shortcut between Alice Springs and the Rock. This back route to Uluru went by the name of the Merinee Loop and it was a reasonably well travelled path, which I guess accounted for its state of disrepair. If I really wanted to shift the blame for this whiplash tour of Central Australia I would probably try and pin it on the bike … it was built for these conditions and insisted on doing autobahn speeds over what was essentially a glorified goat track. However I guess I should fess up to what some shiny suited lawyer would phrase as 'contributory negligence'. I was having the most fun I'd had on the bike since leaving Adelaide weeks before. Consequently, I was doing very little to discourage the bike from taking this low-level fly by of the MacDonnell Ranges.

> Mind you, I couldn't honestly say I had ever heard of anyone successfully 'gluing' a laptop back to good health.

The road reached a new level of viciousness just prior to the top luggage box parting company. A bottomless hole flattening another vertebra to coin thickness made me think maybe I should stop and check everything was lashed on tightly—perhaps even throw on some extra reinforcing straps. I had good reason to be wary about my luggage—the top box had prior form for 'travel independent of the bike'. A year earlier it had broken free while I was riding into the Bungle Bungles. I knew that if I wanted to avoid repeating the Bungle Bungle debacle the box needed additional bracing when the going got rough. Unfortunately though, I was having way too much fun to do something sensible like stop and make with the compression straps.

The one thing I can say about the mistakes I have made in the past is, given the same opportunities again, I can repeat them all. Almost 12 months ago to the day, I was chucking another U-bolt and riding back to the box, which was again lying dusty and battered on the side of a road.

Having bits fall off your bike is never a good way to kick off your day, and I was particularly sensitive about this piece of luggage parting company, as the top box was my reason for being. Inside was all the equipment that enabled me to churn out stories for radio, TV and the Web. The box was my mobile studio. It was home to a DV camera, a bunch of microphones, headphones, a laptop and a heap of other sensitive equipment that most sane people would never strap to the back of a bike. Admittedly the equipment wasn't mine, so that allayed any wallet anxiety, but without my gadgets I was stuffed.

> I fell to my knees in front of the box—it was my little roadside altar and I was praying to the Equipment Gods, begging for kindness.

Riding towards the box I could see it was still closed—this was good. If it had cracked open, everything would have been thrown out and smashed to pieces. The equipment locked inside might still have been pulverised, but at least it would all be in the one place, which meant there was some hope of being able perform a few Super Glue miracles. Mind you, I couldn't honestly say I had ever heard of anyone successfully 'gluing' a laptop back to good health.

I pulled up next to the box, in no rush to jump off and face the carnage. This was one of those clichéd 'heart in the mouth' moments. I felt physically sick about the damage I was going to find inside it. I stared at the box and began arguing with myself, 'Why am I letting the negatives race to mind? I should be thinking, "This road is a war zone and I was going way too hard. I am lucky it was only the box and not me that bounced along the road." I should be thanking my lucky stars that the box is still shut. I might be lucky, everything survived the last time it came off.' After the tirade of self-admonishment I

Another day, another breakage. This time the Mereenie Loop, N.T.

reminded myself, 'You're out here on your own, you have to look for positives and stop being a frigging drama queen, otherwise you won't make it to the end'.

I fell to my knees in front of the box—it was my little roadside altar and I was praying to the Equipment Gods, begging for kindness. I fumbled with the lock, it released and the lid swung up, as opposed to falling off in my hands—a promising start. All the pieces of equipment were still in their allocated nests of padded foam; there were no broken bits loosely floating around. My heart slipped from my mouth and was now resting in the back of my throat. The DV camera was close at hand and arguably the second most fragile piece of equipment in the box. I pulled it from its bay and flicked it on; the motor whirred, the tape locked down and the screen lit up black. 'Black! Oh my god, Black! It's cactus, it's all over…it's…it's because the lens cap is on you idiot!' My heart eased down to the bottom of my throat. I hoped it might slip a little further if I had the same luck with the laptop.

I think it would be fair to say there was nothing in the laptop owner's manual explicitly recommending against carrying the device across tens of thousands of k's of corrugated roads on the back of a motorcycle. Admittedly there was some guff about avoiding excessive shock and vibration, but not once did it mention the word 'motorcycle'. That being the case, I thought strapping it to the back of the bike was well within its operating parameters. Heeding the stuff about 'shock and vibration', I had ordered in some ridiculously expensive rubber to sit the computer on, hoping to insulate it from some of the vibration. This was the moment of truth: I would find out if the couple of hundred bucks blown on this rubber pad had paid off. I pressed the 'on' button and the drive began to wind up. Twenty seconds later the 'Gate's Waltz' blurted out of the speakers and the desktop came to life. My heart returned to its normal seating position.

For a while I remained on my knees, staring at the box of goodies. The Equipment Gods had smiled upon me. Everything still worked and I could keep going, sort of. The mount that held the box on to the bike had snapped, causing the box and me to part company. I gazed at the breakage and concluded that it was best remedied with a fistful of ockie straps. While sizing up 'the fix', I pondered the real reason for the mess I was in. I had been pushing too hard for too long—things were starting to give. The laptop had only just returned from a visit to the PC medicos, now here I was trying to kill it again, not to mention all the other gear. Either I wised up or next time it might be me bouncing along the road, and I wasn't too confident about ockie straps being much use if it came to that.

As I fumbled with the miracle of elastic with hooks, my mind drifted back to the relentless push I was in the midst of, and how it had led to my current predicament. The 'flat strap' travel plan was the product of a timetable–versus–budget equation ruthlessly hammered out back in Adelaide. Unfortunately though, I couldn't blame anyone for the way the sums had worked out, as it was me who had done the maths. Budgets and timetables were just symptoms though; the true cause lay much deeper, and it had to do with an affliction of mine, brought on by desks.

②
How did it come to this?

Every journey begins somewhere. Mine began with my resignation—well, attempted resignation. Resigning then travelling is a habit I've been cultivating for some time. It's an addiction that takes hold after extended periods of being stuck behind a desk. As with all addictions, there is a down side—in my case, it was things like equipment breakages and being stranded in Central Australia. There'd been a chain of desks that had landed me in quandaries like this, and the last link in that chain was in Adelaide. The longer I jockeyed it, the harder it was to deny the urges that were gripping me. I knew the cravings well. I was last seized by them in England.

For several years I lived overseas , working and travelling, and my last stop was Britain. Unfortunately, the Brits grew tired of me, and I was asked to leave (probably something to do with a visa expiry date—although I was also conscious of

Travel by day. Laptop by night. Unfortunately the laptops often came to despise being tortured on the back of the bike. Halfway through Journey 1 vibration took its toll and a quick change in computers was required.

nursing my habit, so I made sure I failed to mention anything about visas when I resigned from my job). Being in no rush to get back to Australia I opted to come home the long way—overland. I decided the best way of turning this into an epic delay tactic was to buy a motorbike and ride the sucker home. It was the obvious and sensible choice: I had never done any motorcycle touring before. Using this same deranged logic, I decided I should record what happened along the way and make some radio about it all. Again, it was the obvious and sensible thing to do as I had never worked in radio before.

With hindsight, going down this path was perhaps not as random as it sounded. And indeed, it may not have been the fault of either my visa nor my habit. I had been working in the back office of a merchant bank in London and after many months I'd come to a crushing conclusion—I wasn't cut out for a banking-type gig. This was a bloody inconvenient realisation, as I had spent a few years studying so I could do exactly this. They say you can't deny your roots and I guess that is what my Merchant Bank Epiphany was all about. My father had been a broadcaster, my brother is a broadcaster, my grandfather had been in the game as a technical person—bloody hell, even my great grandfather had worked as a newspaper journo. I was a marked man. Funny thing is I never knew why any of them did it. All I knew was I had spent a lot of time bumming around the planet meeting amazing people and hearing their stories. I realised that I wanted to tell those stories. I decided what better way of doing that than riding halfway around the planet to gather them?

Through luck, the grace of Vishnu, and every other deity on the subcontinent, I not only survived the odyssey but I also made it all the way home (well almost—I blew the bike up 300 k's from my front door). While recovering from malnutrition and 30 000 k's of white-line fever, I decided to buy a computer and teach myself how to make radio stories. The decision was easier than the doing. Eventually I finished the series and used it to blag my way into the government broadcaster. Not only were they reckless enough to give me a job based on these sketchy travel yarns, but a publisher was also ready to hand over a fistful (about toddler-size) of dollars to put the series on shop shelves.

After weaving my way through a series of ABC positions, the organisation eventually got wise to the trail of chaos I was leaving behind. Action was taken and I was chained to a desk. Ironically, the desk brought clarity. I had joined the organisation to be a storyteller and had been employed as such in various guises (producer, reporter, and so on). Things seemed to be going to plan, until the wheel of fate turned and I found myself back at square one, sitting behind another bloody desk. I resolved to sort it out and decided the best way to do this was to let old vices take hold of me once more. I made some phone calls to float a sketchy travel story idea I had and then wandered into my boss's office and said:

'Ciao, I'm off. My publisher has been gripped by another bout of recklessness and is going to give me some cash to travel around the country and make another radio series. I resign.'

Geoff, my boss, pondered over my offer to clear my desk—something that I considered the offer of the century—and said, 'Give me 48 hours and let's see if we can make this into an ABC project and keep you on the payroll.'

The faith Geoff Heriot placed in me resulted in two years of travel and filing stories for the ABC. Admittedly, it was a little bit more complex than this. In fact, there was a lot of lobbying and selling of my butt to anybody with a chequebook. 'If you've got the cash I can supply some material for you!' was the basis for a lot of spruiking. I pitched too often and asked for too little and the result was that I ended up agreeing to file stories for ABC radio, ABC on-line, and occasionally TV. Considering I have trouble walking and talking at the same time, I was a little concerned about simultaneously supplying material to three different mediums. To complicate it further, I had to cater for multiple audiences. The TV stuff was for school kids around the country. The web and radio stuff had to appeal to everyone from 17-year-olds dialled into Triple J to 80-year-old grannies glued to the ABC Local Radio network. By the time I was finally riding out of town on a spanking new bike, packed to the gunnels with recording and editing equipment I had little idea about how to operate, I was starting to appreciate the quiet desk life I was leaving behind.

3

Not made in Japan

I had envisaged the length of the ride from suburban Adelaide to the edge of the outback to be of an indeterminable distance. After all, where does the outback begin? Fortunately for me, someone had decided that a few k's north of Port Augusta was where it started. To highlight the point they placed a filthy great sign saying so. It seems the debate over where the outback 'begins' and 'ends' was long ago solved by shires and councils around the country. Some decided their patch was far enough away from the big smoke for the 'gateway to the outback' label to be their chief selling point. Other administrations took the opposite view. Terrified of being classified as 'remote', they did as Port Augusta did and placed the sign further up the road and out of view of the town. The curious thing is that if you speak to people who live inside the boundaries declared by these signs they will usually say, 'I don't really consider this to be the outback, mate. You've got to go another 1000 k's before you really get into the outback'.

Exactly where it is, or what is, depends on what pub is being surveyed. No matter how it's defined, the only guaranteed result is that there will be a bunch of people popping veins because they've either been included or excluded. With these thoughts in mind I found the 'welcome' sign outside Port Augusta reassuring. I had been left out of the decision making process and could direct any questions about 'if I was in the outback proper or not' to the local councils.

Sitting on what I was told was the edge of the outback I soon came to a couple of conclusions. Firstly, there was nobody here to talk to, so I probably wasn't going to be doing any interviews with non-outback type people. Secondly, the sun was sitting low in the sky and if I didn't get a move on I would be riding into my planned destination well after Skippy and his mates had climbed out of the sack. Playing 110-kph tag with kangaroos is something I try to avoid.

> I seem to have a bit of a horizon fetish— I like 'em clear and uninterrupted.

North of Port Augusta is one of Australia's many 'Welcome to the Outback' signs. This sign is perhaps one of the better placed ones as nearby the Flinders ranges yield to a vast flat horizon.

Riding off, I began to think the outback sign might have been in the right place. The journey north from Port Augusta sees a transition in the land. The Flinders Ranges peel away to the east and leave behind an endless flat horizon. I seem to have a bit of a horizon fetish—I like 'em clear and uninterrupted. Hills and mountains are okay, but after a while I feel a little boxed in and start to wonder what lies beyond the peaks. Maybe this is why I feel comfortable being in the outback.

Thoughts of my geographical fetishes were soon blown away by a road train. The air blast from the three trailers tugged at my helmet and wrestled with the handlebars. The buffeting finished long before my cursing. When the expletives eventually ran dry I drifted back into the little world within my helmet. Expecting the unexpected is the main rule for motorcyclists who want to grow old. However, for long-distance riders there is another decree: you've got to be happy with hundreds of hours of your own company. The road dictates what sort of company you keep. Long straight bitumen roads are Zen-like environments for the mind to roam. Twisty or corrugated dirt roads are sweat camps for the brain—if the mind strays the bike will follow it over the verge and into the scrub. I knew there was plenty of intense dirt riding ahead, so I enjoyed the meditation of the highway, while also keeping an eye out for the occasional distraction. Hours of uninterrupted road riding can see your mind amble past Zen and into 'date' country.

Fortunately, just at the point when my thoughts were getting a little dark and whiffy, a huge yellow sign loomed up on the side of the road, a multilingual sign proclaiming 'Warning, animals on road' (I was desperate for a distraction). I appreciated the warning but was stuffed if I could see any critters other than those depicted on the sign. The animals on the sign had been used as target practice—which was understandable, considering the lack of living ones to shoot. The painted cow had escaped with a flesh wound but the shotgun hole in the sheep was terminal. As the sign was the only large, upright surface for

L to R: When military interests are involved, it's amazing how far water can be piped. So that Woomera's 6000 desert dwellers could get on with the job of testing rockets and missiles, this lifeline was installed

The only multilingual sign I've seen in two laps of Australia, apparently it's for foreign gem traders who travel to Coober Pedy to buy opal.

500 k's, it was a target for frustrated graffiti artists, and featured spray-can silver script urging 'Listen to Radio Nowhere'. Gunshot wounds, graffiti and confusion over why the only multilingual road sign I had ever seen in Australia should be stuck out in the middle of nowhere made my smelly riding trance appealing, so I punched the start button, re-tuned to 'station nowhere', gave the throttle a twist and continued northwards.

'Australia by road' is a travel package that always incorporates a few long transits down mind-numbingly straight roads. Trundling up the Stuart Highway was my way of paying my dues before the fun started on the dirt. While supposedly you're in the outback, the endless track of tar is a constant reminder that you're safely on one of the slender arteries that join up the country. Keeping the road company was a grey water pipe, snaking its way to the same destination as me—Woomera. In the 1960s Woomera was one of the largest regional centres in Australia, boasting a population of over 6000 people, most of whom were obsessed with flinging rockets into the air. Woomera was chosen for a reason: the desert was a great place to launch large chunks of metal into the sky without having to worry too much about them landing on people's heads. The desert, though, was a bad place to stick 6000 people, largely because it's a little light-on when it comes to water. The military pondered the problem and decided it was easily overcome by laying a pipe to the nearest river—the Murray—which is only about 1000 k's away!

Woomera, for several reasons, was the first stop on my whiplash tour of the outback:
- Firstly, by mandate of the earlier sign, Woomera was technically in the outback.
- Secondly, it was on the way to where I intended to go bush proper.
- Finally, I was fairly confident there would not be too many stockmen kicking around a place like this.

They were all valid reasons and the only thing that was potentially going to change my plans was the waning daylight. The sun was now resting on the horizon and I was still about 100 k's short of Rocket Central. And unless I was looking for roo-skin handlebar warmers, dusk was the wrong time to be riding. On my previous lap of Australia, a bloke in northern WA had told me that you need to shut down and pack it in when there is no longer a two-finger width gap between the setting sun and the horizon. It seemed like good advice, but always left me a little envious of people with fat fingers, as their travelling day was always going to be shorter than mine. Having run the gauntlet of countless sunsets, the 'two finger' theory was firmly on the back-burner; I planned to re-assess this if I ever had a roo knock some sense into me. I did partially heed the advice by stopping to take some photos of an angry sky burning every shade of crimson as the sun dipped below the horizon. The brief photo stop also helped me satisfy another bloke's theory on dusk: 'Wait until the half light has gone, then you've got a chance,' he told me. When exactly 'half light' finished was a debatable point, so I chucked my leg over the bike and pushed on, ready to argue the issue with anything that hopped into my path.

> The land I was riding through appeared to have suffered a vegetation bypass.

The land I was riding through appeared to have suffered a vegetation bypass. To the west I thought I could make out a massive salt lake, but with darkness taking hold it was impossible to really work out what was going on. I decided scanning for wandering animals was probably more useful than playing twilight 'I spy'. If animals did wander out I had one more piece of 'dusk advice' to fall back on. A train driver I met up in the Pilbara had told me, 'Mate, I get cattle, camels, emus, roos, donkeys continually in front of the train. You can't swerve a train and a camel makes a big mess, so what I do is turn the headlights off. It's the light that stuns them. As soon as it's off, their ears tell them to get the hell out of the way.' This piece of advice was the one I liked the most, basically because it avoided the notion of stopping. Grateful as I was for the tip, it wasn't much help because the law dictates that the

headlights of new motorcycles must be wired permanently on. With all forms of advice either useless or ignored, I resorted to the blind hope that no self-respecting animal would think about wandering around such a barren landscape.

With white knuckles and ground teeth I eventually rolled into Pimba, a petrol station accredited a town name, which also doubled as the turn-off for Woomera. All that lay between the two towns was a 2-minute ride and a railway crossing. It was the only railway line I would cross in 1000 k's and I got stuck at the lights waiting for the daily train to go by. Watching the carriages rumble past gave me a few minutes to reflect on how lucky I had been not to have anything jump in front of me—luck that I wasn't confident would hold out if I continued in the same fashion for the next 20 000 k's. I decided that the next day I would stay put, re-group and use the time to chase up a couple of blokes I had heard about who called themselves 'rocket hunters' and spent their spare time out in the desert ferreting around for old crashed rockets.

'When we're on holidays we go rocket hunting', Bruce said.

Bruce was a large, bearded bloke, straightforward and uncomplicated. The simple sentence fell from his mouth as though he was talking about doing something normal, like going on the annual family fishing trip. Poking around thousands of k's of desert, trying to find old crashed rockets does have some

L to R: Lake Hart, a miniscule part of Woomera Rocket range Len Beadell's grader, the workhorse used to open up Central Australia by carving out around 6000 k's of roads. The road network was needed so the folks from Woomera could retrieve their rockets.

parallels with trying to bag an elusive fish. However, no sane person would consider such an angling expedition unless they had a bit of inside info. Bruce did—he worked as the Rocket Range Safety Officer, a gig that gave him access to the trajectory details of all the rockets ever launched at the range.

'If we included all the 3-inch and 5-inch test rockets that were used for tracking, we're probably looking at about 10 000 rockets launched since 1947', he told me.

The numbers came easily to Bruce, as he'd been living and working at Woomera for about 30 years. Long enough, you would think, to make a fellow want to get as far away from rockets as possible on his holidays.

It is not every day you meet a bloke who calls himself a 'rocket hunter', so I made sure I had my recorder running, thinking it might make a half-decent radio yarn. In my usual balls-up interviewing style, which attempts to put people at ease through the use of flippancy (and usually achieves the opposite), I asked, 'What do your mates say when you tell them you're going out rocket hunting? Do they look at you as though you've got a hole in your head?'

In Bruce's job he dealt with dickheads regularly, so he was unfazed by the question.

In theory there were not supposed to be any rockets left to retrieve. A clean range policy dictated that what came down must be picked up.

'They ask if they can come,' he replied.

In theory there were not supposed to be any rockets left to retrieve. A clean range policy dictated that what came down must be picked up. The problem with the 'pick-up' theory was the distances involved. Most of the early rockets were sent thousands of k's out to northern Western Australia, lobbing into a potential landing area of around 2.5 million square k's. Back in the 1950s much of this area had avoided the gaze of white fellas, so there was a distinct lack of roads facilitating easy rocket retrieval. Nonetheless, the rockets had to be collected, so a bulldozer and a grader were handed over to a bloke by the name of Len Beadell, who, with a few mates, set about gouging out 6000 k's of roads through the centre and west of the country.

> 'The attitude of the day was that there was a lot of nothingness out in the Simpson, so the rocket was just left there', Bruce said.

This was all well and good and Len became a bit of an outback exploring legend in the process. The problem was, though, that not all of the rockets headed north-west. The Redstone Rocket, which carried Australia's first satellite into space, was launched northwards, out over the Simpson Desert. The launch was a success and the first stage of the rocket came to rest in the Simpson, a spot Len and his bulldozer did not visit.

'The attitude of the day was that there was a lot of nothingness out in the Simpson, so the rocket was just left there', Bruce said.

Bruce and his mates considered the Redstone wreck worth hunting for. After all, it briefly put Australia in the space race as one of only 4 nations in the world to have launched a satellite at that time. (The satellite may have been ours but the rocket was an American leftover. They bequeathed us a spare missile so we could have a crack at strapping something to the top of it and flinging it into orbit. Who owned and built what was a technical point, which at the time did not stop us from saying we were part of the space race.) Fine print aside, the rocket was part of our history and it was rotting out in the desert. Bruce and a few mates set about finding it and bringing it back home to Woomera.

Retrieving the rocket required a little more than just a trailer hooked onto the back of a ute. A six-wheel-drive truck was conned out of the army and a carefully planned mission to 'find and retrieve' was set in motion. Fortunately Bruce and his mates were no outback virgins and venturing into the Simpson was well within their skill set. Not only did they have bush experience but also, thanks to the trajectory records kept at Woomera, they had a reasonable idea about where to start looking. This was a good thing, otherwise the rocket would have been the proverbial needle hidden among 170 000 square k's of inhospitable desert.

'The trick is, if you abuse it, it will bite you on the bum when you least expect it and you will end up being a modern day Burke and Wills. If you go into that country without adequate spares, food and water: (a) you're an idiot and you shouldn't be there in the first place, and (b) you'll get caught sooner or later.' Bruce said this in a way that seemed to be particularly relevant to me.

'So my mission should be stick to main tracks', I blurted out as I chucked my leg over the bike. Bruce agreed enthusiastically and watched me ride off past Rocket Park and the Redstone remains.

Passing through the vacant streets of Woomera I felt conspicuous in the emptiness. The former desert metropolis was well on the way to becoming a ghost town, the bulk of the residents having moved out with the last of the rockets, leaving behind street after street of vacant houses. In the centre of town were more vacant buildings: large, empty, besser brick accommodation blocks that were once owned by the Americans and sold recently to the Australian government for the princely sum of $1. There was just way too much infrastructure for so few people. With the absence of people the desert had begun to make a claim on the town— emus now wandered the streets.

Emus picking their way down a main street might be amusing to the average Australian, but it was utterly bewildering for the 50 or so Japanese scientists who were also living in the town. The town recently had a temporary population booster shot, with the Japanese coming to trial a model of what they hoped would be the next generation of supersonic planes.

Supersonic jets, like 'rocket hunters', seemed to be missing from all the marketing bollocks I had consumed on what the outback was all about. With that in mind, plus the opportunity to check out some modern day Chuck Yeager-type gadgets, I headed for Mission Control.

The Japanese test site was located about 50 k's out of town, a short journey that soon reinforced the idea that Woomera, even in its diminished state, still existed to serve a fully functioning test range. The first reminder came about 10 k's out of town where a sign warned, 'Restricted Area, Turn Back', and the road beyond was blocked by an electronic boom gate. Adjacent to the boom was a sentry box, long ago abandoned, with the personnel now superseded by an intercom and security camera. The intercom asked me to identify myself and give the bike registration number. A short silence followed and the boom began to rise. I assumed someone was satisfied that I was meant to be there. I mumbled some thanks, closed my visor and rode on to the range.

The Woomera test area was well chosen. The site was located on an endless expanse of barren ground. Stretching out into this gaping landscape was a flat, straight bitumen road—a mix of driving conditions that was the undoing of some of the car-driving Japanese. The scientists came from a congested country where getting a car into top gear is nothing short of a fantasy, so they embraced the open road freedom with a zeal that led to the car rental company suffering a stock shortage—not because everyone was renting a car, but because the cars that were being rented kept on being trashed. Apparently the problems lay with speed, swerving to avoid animals, speed, failing to negotiate a couple of corners, and speed. While riding out to the test facility I was hoping for two things. Firstly, that I would not encounter any scientists in cars; and secondly, that they would have more luck controlling their plane than their cars. (I found out months later that unfortunately they didn't. Their first test flight slammed into the ground at a spectacular—you guessed it—speed.)

> It all looked worthy of a plot to take over the world, but I was assured it was focused purely on developing ridiculously fast passenger jets.

Forty k's later I arrived intact and feeling as though I was in the midst of something worthy of Commander Bond. Teams of Japanese dressed identically in pale blue uniforms scurried around the desert clutching clipboards. The object of their attention was a large refrigerated hangar. In

Exactly what you would expect to find on a Rocket Testing Range. A movable, air-conditioned hangar, housing a rocket with a model plane strapped to the back of it.

keeping with the 007 theme the hangar had been built on rails, and at the appropriate time it could be wheeled out of the way to allow the crouching rocket launcher hidden inside to stretch out into launch position. It all looked worthy of a plot to take over the world, but I was assured it was focused purely on developing ridiculously fast passenger jets.

I'm the first to stick my hand up and say I am an ignorant bugger, especially when it comes to aeronautics. Considering, though, that they were working on a project involving a supersonic plane, I was a little surprised to see a rocket. (Then again, I was on a rocket range, so I guess a plane would have been just as big a surprise.)

A perfectly groomed man called Dr Sakata was overseeing the small blue army. His crisp white shirt and perfectly pressed blue suit seemed to complement the warehoused world of modern precision: a techno sanctuary shielded from the ancient desert by air-conditioning and insulated walls. The scientists had come to Australia to conduct field tests on the aerodynamics of their design. 'Design' was the key word in all this as they were lacking the real plane. In fact, it hadn't even been built. So instead they brought with them a scale model, eleven times smaller than the final passenger carrying version. Because it was a model it lacked the grunt required to get it up to Mach 2—in fact, it lacked any engines at all. So a compromise was hatched, which involved strapping the model to a rocket and hurling it into the atmosphere at a ludicrous speed. Once the rocket was up to battle speed they would detach the model and plunge it into a nose dive

Rocket, and plane, in 'action position' after the hangar is towed out of the way.

until it broke the sound barrier a couple of times over. With the model near terminal velocity the data recording would then begin, lasting for a grand total of 40 seconds!

Testing the plane back in Japan was a possibility, but the problem was it had to be done over a stretch of water. The water didn't pose any difficulties but the fishermen on it apparently did. The tests would require clearing out all the fishermen for a day, a process that Dr Sakata assured me would be very expensive, because of the compensation that would have to be paid. In its heyday, Woomera had faced the problem of making sure the flight path was clear. The people at risk in those days were pastoralists, and the testers had to ensure they were out of harm's way when rockets were launched. Although this solution didn't require compensation, it did involve building concrete bunkers for all the affected farmers, giving them a safe haven to shelter in during tests. Apparently the farmers thought the bunkers were great, mainly because they could put chairs on the roof and watch the rockets fly overhead. Japan's Aerospace Laboratory opted to avoid such primary producer problems in their homeland and the test team found itself in the outback.

'The landscape, completely different. There's nothing, except sand, small stones and small grasses.' Dr Sakata's words were delivered slowly and deliberately, taking long pauses before blurting out his final conclusion.

'Very nice for testing …' (pause) '… not very nice for the living.' He laughed at his joke, a 'bah ha ha' that sounded like it belonged to what I would have once considered a bad Japanese impersonator.

He continued: 'For me it is the first time to see nothing for 180 degrees, a clear horizon, it is very strange'.

It was all a little strange for me as well, except in my case it was coming to terms with the countless dollars worth of equipment and truckloads of technology. Japanese technology appeared to be something Dr Sakata was very proud of. As we walked in the direction of where my motorbike was parked he made what was perhaps a reasonable assumption, 'So you are motorcycling, you must be on a Japanese motorcycle?'

'No', I responded, taking strange delight in the fact it did not come from Japan. He seemed a little surprised and maybe even a little crestfallen.

'Oh, it's so big. How do you manage?', he said when he laid eyes on my bike.

'Damn right it's big', I thought churlishly. For some reason the Japanese tech-fest had irked me. I hid this strange spike of spite by assuring him that all my cameras, microphones, computer and so on were Japanese. He seemed to accept the compensation, but it was hard to tell. I suspected that from his vantage point of 5 foot nothing the concept of wrangling with a bike so large, so un-Japanese, was a little hard to comprehend. I was being childish and I resented myself for it, especially as he had been so friendly and giving with his time. I bade him farewell and nearly dropped the bike as I climbed on to it, possibly karma or clumsiness attempting to put me back in my place.

An ancient radar station—Australia's technical contribution to the Japanese quest to build commercial supersonic jets. I think the deal was 'You can use our test range, but for safety's sake we are going to keep a very close eye on things with this 1950's radar desk'.

I rode off and tried to rationalise why my nose had got out of joint. While I was amazed at all their technology, it was also the source of my consternation. Sticking in the back of my throat was the fact that Australia's only contributions to the endeavour were some dilapidated old buildings and 100 000 square k's of open space. I lie, we did contribute some technology: an ancient radar, housed in what was essentially a shed down the road. The glory days of when we viewed ourselves as being part of the space race were long gone. I tried to rationalise this by musing that at least we turn out a lot more sports stars than we used to. Sure, that might not do much for the GDP but it sure as hell makes us feel good on the weekends. The argument over our place in the world raged on in my head while the middle man up there beckoned me to put my earphones in, crank up some tunes and drown out the bloody row.

It's got nothing to do with pasta

Like most of my departures, leaving Woomera didn't go to plan. Usually I run a little late, but 3 days late was a new record. It seemed it wasn't just technology that the Japanese brought to town. They had also brought the latest flu strain. Being sick as a dog had some flow-ons—my laptop decided to go out in sympathy. The screen turned up its toes and began projecting in one shade of black. No computer equals no output. This meant the people who had funded my story-telling expedition were going to start asking awkward questions. It was all going to plan—sick, 3 days late and a major breakage!

I wasn't surprised to find out that Woomera was a little light-on when it came to computer repair shops, so the only solution was to do the 'bush mechanic' thing. As I began attacking the laptop with a screwdriver I pondered on how desperation can create an instant Mr Fix-It, no matter how poorly equipped on the knowledge front. Good fortune was on my side, and I stumbled across the fault: a damaged cable, eaten away by the vibrations caused by tens of thousands of k's endured during the previous year's trip. The fault was intermittent, but became permanent when I poked and prodded and the ribbon cable tore through. When a calamity this big happens a strange calm settles over me. It's the small-scale stuff I ulcerate over. The only solution was to get it to Sydney for repairs. A good plan, but overnight delivery services were as common as computer repair shops, and even if they had existed, it was a Friday night, so I was still stuffed.

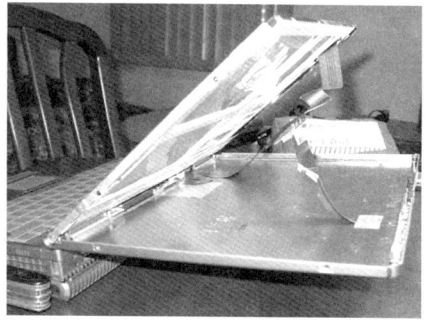

My computer was my office; inconveniently it decided to have a mild coronary in Woomera. I was stuffed without it, so the only option was a frantic attempt at CPR. Sadly opening up the back of the laptop proved fruitless; in fact my tampering escalated its position on the critical list and I ended up having to ship it out to a specialist.

A few questions around town pointed me in the direction of a bloke who was flying out of Woomera the next day. The following morning the laptop was winging its way to Adelaide. Thanks to a few miracles, and some people I promised a lot of beer to, it was on a desk in Sydney by Monday morning with the repairs in progress. What all this proved was that I was a lucky bugger and I should be more careful with my equipment. It also proved that I was anything but isolated.

> When a calamity this big happens a strange calm settles over me. It's the small-scale stuff I ulcerate over.

Without my computer I felt kind of naked. However, there was an upside. I had spare time on my hands, time that could be spent doing bar research. From what I heard, Coober Pedy—my next destination—was a great place to do exactly that.

Roadside distractions for the Coober Pedy bound.

Not far out of Woomera I rode past the glaring white expanse of Lake Hart. Why tracts of land that are saltpans 99% of the time deserve the title of 'lake' escapes me. Perhaps early explorers were just being optimistic, trying to convince themselves they had stumbled across it in a bad year, though I doubted that anyone could have such a degree of optimism when surrounded by landscape so dry and baked.

North from the lake the road droned on through flat and featureless country, an endless horizon only occasionally interrupted by signs declaring that Marla Roadhouse was now another 100 k's closer. After a few more hours and a couple more signs promising a range of roadhouse excitements like 'licensed restaurant, showers and fresh vegetables', my excitement over Marla was redlining and my pulse needed to be calmed. Glendambo came to the rescue.

The sign out the front pretty much summed up what to expect from the roadhouse village—that is, a place to fill the tank, grab some nosh, vent the bladder, maybe chat to another traveller (mind you, they aren't as friendly on the main drags), saddle up and keep going, in my case a couple of hundred k's more to Coober Pedy.

A few hours later, large pictorial signs illustrating people falling into bloody great holes heralded the start of opal country. The holes, and any falling

people, were obscured behind
the large mounds of heaped white earth, an acne of tailings that stretched across the face of the mining area. The temptation to take a side road and explore among the mounds was quelled by another sign, this time warning of a short life expectancy for inexperienced people who entered the mining area.

As I continued to ride past the fields I caught occasional glimpses of mining rigs parked among the mounds. They were the only indication that anyone actually took the risk of entering the fields. All the trucks were rusted and battered, and each had a crane-like arm jutting out at the back. The arms were obviously not for lifting, as welded to the end of each of them was a large drum. The few rigs that were running spewed out plumes of dust and burnt diesel, but I was left none the wiser about why a large drum suspended at the end of a crane arm was crucial to digging opal out of the ground. I wasn't sure if I cared about it either, as these rigs were part of traditional mining set-ups and I was on the hunt for a bloke involved in a different mining method that went by the name of 'noodling'.

'It's got nothing to do with pasta', laughed Mark as he shoved another beer my way. 'In fact, I don't know where the name come from really.'

The irony that Mark ran arguably the largest noodling business in the region but couldn't tell me where the word came from appealed to me. My fixation with etymology was forced to the back of my mind when I began

27

'It's a bit like a seagull picking through the scraps.'

another bout of 'beer cap wrestling'. The bottle tops lacked the 'twist' function I understood so well; instead they required a DIY opener. In Mark's workshop this equated to a blunt piece of steel, which was used as a lever between your thumb and the cap. It was clearly designed to sort the boys out from the girls and I was beginning to look like I needed a skirt. After another protracted bout with a reluctant cap, the 'opener' had more of my flesh hanging from it. I watched my blood trickle down the side of the bottle and wondered if getting opal out of the ground by noodling was half as painful as extracting beer from Mark's bottles.

Noodling involves going through the slagheaps of old opal mines and picking out the gems the original miners missed, or in Marks words:

'It's a bit like a seagull picking through the scraps.'

To find the stones previously missed, noodlers resort to a bit of basic technology—a black light. When opal is passed under a black light it gives off an iridescent glow, making it extremely easy to pick out. There's one catch to this process—it has to be completely dark for the black light to work. The opal fields of Coober Pedy are the last place in the world you would choose to crawl around in the dark and do a bit of black lighting. One estimate is that in the 5000 square k's of mining area there are around a quarter of a million shafts. I doubted the number, but whatever the statistic, Mark had some basic advice: 'It's not a place to shitlair around in'. Or to spell it out:

- never run,
- never take a step backwards, and
- never go into the fields at night.

With night noodling ruled out, the only option was to turn day into night, which requires a darkroom. Most noodling darkrooms are glorified sheds on wheels, owned and operated by a single person who spends a lot of time running around doing the following:

- attacking the talc-white tailing mounds with a bobcat and scooping up dirt and rocks,
- emptying the rubble into a crusher,

- breaking up the rocks with the crusher and feeding them onto a conveyor belt,
- cranking up the belt and running into the darkroom, and
- frantically scouring the conveyor belt for glowing bits as rock zooms under the black light.

The cycle doesn't last very long and the noodlers find themselves quickly back in the bobcat position. This is undoubtedly a relief, as most of the black rooms are not air-conditioned and Coober Pedy experiences evil temperatures during summer. Forty degrees is an average summer day, and inside a poky noodling hut it's at least 10 degrees hotter.

Mark decided that doing laps between the bobcat and the 'mobile shed' was all a little mickey mouse; he figured that with a few more people, and some bigger toys, the process could be streamlined. The bobcat was replaced with a filthy great earthmover that focused on ensuring the crusher never ran dry. Mark's darkroom was more of a refrigerated semi-trailer than a shed, with 3 conveyor belts whipping through it, each continuously monitored by a dedicated opal spotter studying what was being carried on each one. They were big toys that required just as big dollars.

'There are still some good luck stories around Coober Pedy. Some people have a fair run, some people don't, but if you stay in it long enough we all get to have a go. It's just like buying a Cross Lotto ticket,' Mark said laughing.

I suspected the odds were now much more in Mark's favour, especially since he had progressed from picking through old slagheaps to a process that was now essentially open-cut opal mining. Mark hurled me another beer—and the cursed opener—before he continued:

'You've got to find your own dirt, otherwise we have to wait for the miners to dig it up…and the miners don't seem to be turning up. They're getting too old. And the young blokes don't seem to be as keen.'

By the time he got through the sentence the opener had claimed another chunk from my hand. The Ying and Yang of alcohol was coming into play: the Ying being that my judgment was fading, so the number of divots in my hand was increasing; the Yang being that the alcohol was dulling the sting in my hand. Worrying about some paltry nicks on my hand was probably not the tough thing to do while speaking to a miner, so I re-focused on the conversation.

Mark went on to explain that many of the original miners had come to Coober Pedy after working on the Snowy Mountains scheme in the 1950s. In Mark's opinion these were blokes used to 'fair dinkum hard work'. The 'Snowys' background probably also explained why there was a Greek club, a Serb Club, a Hungarian club, an Italian club, etc. around every second corner in town.

In all, there were 45 nationalities making up the population of 3000. The 3000 figure surprises a lot of people, basically because there doesn't appear to be enough houses for them—which there aren't, as half the population have dug burrows and live underground.

With many of the Snowy veterans now getting too old to dig, the region needs an influx of young miners to replace them. Mark said the ones who do have a crack soon pack it in because, in his books, 'there are easier ways of making a dollar'.

With the old-time miners dying out Mark reckons there is change in the air.

'It will go like fishing and farming. It's getting too hard for the little people so it will end up with a few big companies running the show.'

I bade Mark farewell and wandered off thinking that he was probably in the running to be one of those future companies. As I stumbled through the streets of Coober Pedy, nursing a belly full of beer and a bloody hand, I took some comfort in the thought of a local being at the forefront of where the industry was heading. The 'frontier town' feeling of Coober Pedy would be lost if corporate interests took a liking to the place.

With the sun setting I decided to scamper up a ridge and watch it go down (scamper was a little generous, it was more a series of upwards falls). Besides checking out a wicked sunset it also gave me a chance to suss out the chimneys that stick out of the hills around town. 'Chimney' is perhaps not the correct term: 'exhaust vent' or 'breather hole' is probably more in the ballpark. The vents, each topped with a pointy cap, poke up from the underground houses. Supposedly they

Ventilation tubes poking up from people's homes in the warren that is Coober Pedy.

help with air circulation, although I suspected they might also provide a great opportunity to tune into the domestic turmoil of the homes below. Sadly the reception was non-existent, so my beer-addled brain moved on to how much fun I could have by dropping a few fire crackers down the chimneys, posting a little bit of chaos directly into people's living rooms. The fun, though, had to stay in the fantasy realm, as unfortunately I was unarmed. In hindsight that was probably a good thing. Coober Pedy has a history of being a volatile town with a reputation for meting out its own justice. It's a reputation that probably goes some way to explaining the frontier feeling still present.

Tourism has changed the place a little, with a few flash hotels popping up to grab the travel dollar, but other than that there didn't appear to be any concerted effort to tart up the place. Road edges blurred as they cambered away under the shroud of dust that coats the town. Utility poles sprang up randomly from baked ground that had a topsoil bypass. The wires the poles supported crisscrossed the town in random patterns that betrayed any notion of planning. It's perhaps one of the few towns in Australia that could be thought to belong to another continent. The dust, the raw earth and the lethargy that comes from the extreme heat had me toying with thoughts of previous journeys to places like Turkey.

I rolled down the hill and made for my burrow, bed being the ultimate objective. My head began a dull throb as I nursed it to the pillow and buried deep into it, partly hoping to cushion the ache, partly trying to hide from thoughts of riding the next day with a hangover and a hand festering inside a motorcycle glove. Sleep, unfortunately, did not come easily. While underground living is prized because of the constant temperature, the temperature was a little too much on the warm side of constant for my nocturnal desires. However, I was eventually lured to sleep by the silence that was as absolute as the blackness.

Packing the bike the following morning was a sluggish process. The 'beer cheer' had been replaced with a slow fogginess, hampered further by a tender and scabby hand. My stomach was hankering for a greasy breakfast, the rest of me was begging to hide in a cool dark corner. The concept of throwing a leg over the bike and pushing further north was something I was not amped about.

I could only drag out the packing for so long, as I had arranged to visit Mark's noodling machine and was due there by morning smoko. I had thought out my day so well—ride north for 50 k's and check out the mining operation, then ride back south 50 k's so I could hop on a phone and get the verdict about the health of the laptop. A hundred k's of riding so I could be in exactly the same place as when I started! The progress report on the laptop would determine how much time I had up my sleeve. If the laptop was still ill I would have time to chuck a right out of Coober Pedy and do the Oodnadatta track. If the laptop had been revived I would go straight to Alice Springs to meet it—I had to start catching up on the work I'd been promising people.

With my choices mapped out, and 50 k's under my belt, I stood in the giant bulldust pit that was Mark's noodling operation. If Mark had told me it was a talcum powder mine I would not have questioned him. The dust was ultra fine and quickly filtered through my riding gear. Walking onto the site set off small dust explosions that billowed up from each foot placement. The powder was shin deep and my boots only made brief guest appearances between each step. As I waded towards the noodling machine my retinas were assaulted with a glare usually reserved for the snowfields. However, there was no alpine cold, just searing desert

The process of scrounging through old mining tailings for missed bits of opal is known as 'noodling'. It is usually done on a small scale, at this magnitude I think it is probably more accurate to call it 'open cut opal mining'.

heat and the flies swarming to dine on my sweat. Nature was conspiring to add discomfort beyond what I was feeling in my eyes. Near the noodling hut the swarm of winged maggots remained unperturbed by the deafening crescendo of rocks being pulverised and a generator screaming. I was keen to get into the darkroom and escape the hostilities of the pit.

From the outside the noodling room looked like an industrial coolroom. The walls were made from insulated aluminium panels, enamelled white and perfect for blending in with the mine. The door was a commercial fridge door, held tightly closed with thick vacuum seals. Entering the room is done quickly; firstly because daylight cancels out the black lights, making it easy for passing chunks of opal to escape the hands of the pickers. Secondly, with the heat being so intense, the pickers aren't too keen on letting the air-conditioned air escape. The final reason is so the room can re-pressurise—air is pumped in to keep dust to a minimum.

Comparative quietness and black coolness enveloped me as I stepped through the door. My eyes slowly adjusted to the black lights and I could make out three conveyor belts speeding from left to right, each carrying an endless cargo of rubble. Among the drab stream of rocks were occasional iridescent flashes that instantly attracted the hands of the pickers. If the rocks were deemed to be worthy of a second look they were hurled into bins on the floor. The bins looked like they contained a fortune, but I was assured that not everything picked was of value. The final assessment would be made later, after the rocks had been washed and cleaned.

Brett, the picker on the belt closest to the door, began telling me he'd been working as a picker for three years and was used to the conditions. He said the only aspect of the job that had changed in that time was that he now wore blue-tinted glasses, as they helped cut out glare from the black light, making it easier for him to spot the opal. I guessed they also shielded his eyes from dust, but that didn't seem like manly miner talk at the time, so I didn't raise it.

While talking to me, Brett's eyes remained fixed on the stream of rock hurtling under his nose. It was an intense gig in gruelling conditions, which probably explained the length of their working day.

'We start at 7, have half an hour smoko at 10 and then wrap it up at 12.30. That's when the second shift starts. I don't reckon I could do an 8-hour day—much longer than 5 and you start to get nauseous', Brett said.

I was finding out what he meant. Continually staring down at a conveyor belt fanging past was a great way to bring on nausea. Apparently this is exacerbated further by (a) lack of familiarity with the black lights (b) hangovers. Being guilty on both fronts I was starting to feel a little queasy. Just in case I had any doubts about the symptoms Brett continued to elaborate:

'You get the head spins and you feel like you're going to throw up so you've got to leave the darkroom pretty quickly, 'cause you don't want to spray the person next to you.'

It was a fair point, so I quickly said my thanks and bolted for the door. Outside, my mind was distracted from my stomach by my retinas protesting about the jump from extreme darkness into extreme whiteness. The stinging lasted long enough for my breakfast to do a U-bolt and begin sliding back to where I wanted it.

I made my way to the bike and began the ride back to Coober Pedy. As I passed the occasional miner labouring in the boiling barren fields I reflected on Mark's words about there being easier ways to make a dollar than opals. I kind of agreed, even if it meant I was tied to the whims of a laptop.

Three hours and 100 k's later I was again riding past the noodling pit. The phone calls in Coober Pedy had ruled out the Oodnadatta track option—the laptop would be in Alice Springs in a few days. I had to meet it as it came off the plane so I could start catching up immediately. Two weeks on the road and I was yet to file a story. Worse still, I was yet to get off the flaming bitumen.

One of an estimated 250 000 reasons why you never step backwards in the Coober Pedy opal fields.

Mincemeat kneecaps

The only things I saw for several days were the inside walls of the hotel room-cum-temporary production suite. I was locked into story mode. A couple of radio stories were bashed out, a stack of stuff for the website, and plans laid for a TV yarn further up the road. After a few days of being glued to the laptop my initial joy about our reunion had evaporated. Cabin fever was on the rise and a distraction was needed: it was found in the form of the Finke Desert Race.

Outback races traditionally involve moleskins, beer, stockman's hats, beer and tomorrow's dog meat being thrashed around a dirt track by short blokes in lairy threads. The Finke Desert Race is a little different. Firstly, there are no horses involved; secondly, and perhaps more importantly, moleskins are also in severely depleted numbers. The Finke Desert Race is essentially for loons, or at the very least people who have had a fear bypass. The race involves cars, buggies and motorcycles blasting along an old railway service track between Alice Springs and the town of Finke, a few hundred k's south. It sounds simple enough, but the sting is in the preparation of the track—there is none. Indeed the race organisers pride themselves on the fact that there has never been any maintenance done on the track. Countless years of abuse, dealt out by hundreds of dirt bikes, buggies and 4WDs have seen the corrugations on the track grow from garden variety teeth rattlers to metre high spine crushers. I'd heard the best way to deal with them was to skim across the top—a feat that required a speed of around 140 kph! The race sounded like there was a lot of potential for pain. This, combined with the moleskin exclusion zone, meant it warranted a look, so I set about tracking down some racers.

Alice Springs by air...or more accurately from Mt Gillen.

'Ride towards the airport. When you are nearly there, look to your right for a starting tower in the middle of the scrub, that's where we will be', said Andre, and hung up the phone.

With these vague directions I headed out into the desert. The plan was to join Andre and his mate Nate who were going out for a practice session, and in the process I would find out a bit about the race and also ride the course with a couple of blokes I feared had a tenuous grip on sanity. The latter thought, however, didn't really add up. I had already met Andre in Alice the day before, he fixed a broken luggage lock for me, and at the time he came across as one of the most calm and sane people I had ever met. I concluded that it must be a Jekyll-and-Hyde motorcycle thing and fully expected to see him white-knuckled and drooling when I eventually found him out in the desert.

An hour after Andre's call I was staring at a scrubby horizon, but unfortunately no tower. In the distance, hurtling across the desert towards me, was a streak of billowing dust. As the cloud grew nearer it was joined by the scream of an engine. In the time it took me to think, 'it must be Andre', a motorcycle had blurred past, leaving me open-mouthed and choking on the dust fallout. Between coughs I figured, 'follow the cloud and you'll find the tower'.

Standing under the tower was Andre, sans drool and distorted knuckles. In fact he looked like he had the previous day: calm and relaxed, albeit a little dustier. I was introduced to his mate Nate and a plan was laid out—follow them. It sounded straightforward, but having just witnessed Andre's low-level fly by my gut tightened at the thought of the possible speeds involved. I began to question myself as to why I had agreed to go on a ride with them.

Twenty minutes later I was again asking the same question. The going was ridiculously hard and hot. I was cooking in my riding gear and fearing spontaneous combustion. On the upside, my fear about speed was unwarranted because the first section of the track was apparently not the fast bit. In fact, it was the opposite—kilometre after kilometre of tight twisting sand trails. My bike was nearly twice the weight of their race bikes and I wallowed around the track. I didn't expect it to be easy, but certainly not so bloody hard. Even if I was on a bike built for the terrain, there was still no way I could contemplate flitting over the whooped-out landscape at mach 2—I just didn't have the skills or the goolies. I need not have fretted, as apparently not everyone is successful at it. In fact success is so tenuous that the organisers have three rescue helicopters in the air during the race. With long distances and the lousy roads it's the only way to quickly evacuate injured competitors.

Sensing my own existence nearing a marginal state, I called for a stop to the sand torture by using the pretext of needing to take some photos (I may have been dying, but admitting it just isn't the blokey thing to do, especially when there are bikes involved!). While fumbling for the camera, and in between gasps for air, I conveyed to them what sort of photos I wanted to get.

'Fast ones', I panted, and they rode off.

I'm the first to admit that I'm 'photographically challenged', but trying to catch a snap of these guys zipping past at sub-light speed proved to be a bigger test of my limited skills than I'd bargained for. Sheepishly I asked them to do a few more runs.

'I want to try and get a few different framings', I shouted over their idling bikes. It was part of the truth; the other part was that I wanted to actually get them in the frame, as opposed to the first attempt, which featured only a back wheel disappearing out of the picture.

Fink Desert Race practice, a rite of passage for Alice Springs riders and any other loons who can work out how to fit Warp Drive to a dirt bike.

When they got back I duly told them they were insane. That's a compliment any red-blooded bloke loves to hear about their riding ability. The difference was that I was using the term in its truest medical sense. The speeds these blokes were doing begged questions about how well their synapses were wired up. Sure, the speed was the dangerous part, but if their health insurers could see they were doing it in dust and subsequently near zero visibility, I would expect small countries would be demanded as premium payments. I asked Andre if he could see the track at all.

'You ride by braille', he said simply. 'You don't actually see the ground you're going over. All you do is watch the fella's helmet in front of you. If that disappears then you either brace yourself to ride over the top of him or you relax because he has shot off into the bush.'

I absorbed this and thought, 'Not only do you have to worry about the damage you can do to yourself, you also need to factor in the

> 'You ride by braille', he said simply. 'You don't actually see the ground you're going over.'

possibility of being used as part of the track by your fellow racers'. But I kept these deliberations to myself. Andre was probably already thinking that I couldn't ride—there was no reason to add 'blouse' to his list of my shortcomings.

The carnage on race day is dealt with first-hand by support staff dotted all along the track. If an accident is serious, the rider gets a chopper ride to the Alice hospital. However, when riders go out for a bit of practice, evacuation is their own responsibility, so it's best to ride with a mate and that's

Some of the staff of ABC Alice Springs insisted on proving to me that Outback Australia wasn't entirely flat. After an hour of being frog-marched up a filthy great hill called Mount Gillen, I was forced to agree with Ian, Ingrid and Samantha.

the reason Andre and Nate were out there together. Accidents in practice are common: the weekend before our outing a bloke had failed to successfully navigate his way past a tree. That earned him a trip to Adelaide to get two broken arms repaired. In another practice accident, a rider failed to negotiate a bend and ended up in the Alice hospital with a broken leg. With casualties being so high and Andre being married with two kids, I wanted to know what his wife said about it all.

'Be careful dear.'

'Really, she doesn't get too stressed?' I asked him.

'Nah, I just say goodbye and I'll see you in hospital.'

Andre tried to reassure me he did look out for himself, but I wasn't entirely convinced, especially after he told me his knees were cactus from riding.

'Occasionally, if you misjudge something, like the speed you're doing when you charge into a dry creek bed, the only way of making sure you don't get flung off the bike is by jamming your knees under the handlebars', Andre said through dust-caked lips.

Apparently the only downside to regularly aborting these high-speed ejections is ending up with mincemeat instead of kneecaps.

With so much pain involved, I began to wonder if doing the Finke was more a rite of passage than a race.

'Yeah, pretty well. It's a hell of a race. A lot of people don't do it because they don't think they are good enough. You don't have to be good, you've got 3 hours to get there and it's only 230 k's', Andre said.

It might only be 230 k's, but if you wanted to tackle it in the average 4WD and still have some resale value left at the end, it would probably take about 8 or 9 hours. The thought of competing in it seemed well beyond my dreams. Andre, however, offered my aspirations a lifeline when he said:

'We've just got to get the postie bike class happening for all the old blokes.'

Leaving Andre and Nate to get on with some serious speed, I rode off dreaming of glory on a bike one-tenth the capacity of my current steed. I was disappointed I wouldn't be around for race day, but that was still a couple of months away and I was due to leave town the next morning. I also knew that I didn't need to watch other riders struggle, as I had my own endurance ride ahead of me and it was probably going to be the hardest part of the journey. I was soon to head west to Kalgoorlie—all that lay between me and it was about 2000 k's of desert.

6

Bouncing bombs

I cut out of Alice Springs via Albert Namatjira country, the MacDonnell Ranges. The hilly terrain, the abundance of grass, trees and the colour 'green' was the total opposite of the flat barren landscape I had expected. (Which also highlighted the fact that I could recall 'bugger all' of Namatjira's paintings.) It was a landscape completely contrary to the dry barren vistas I had left behind in northern South Australia. To reinforce the point there was even water in the river adjacent to Hermannsburg, my first fuel stop for the day. In town I commented excitedly to the bloke running the general store/service station about how healthy it all looked. He was quick to point out that I was seeing the exception and not the rule; apparently it was thanks to three good years of rain.

The west MacDonnell Ranges, Namatjira's inspiration and also part of the back route to The Rock.

Besides giving meteorological summaries, the store manager also doled out the permits required for the Mereenie Loop, the back route I was taking to Uluru. I thought the permit was purely to get permission to travel through tribal lands, however the manager said it was more about safety—a way of keeping track of the vehicles out on the route. A safety procedure like this made me wonder what state of disrepair the track was in and I began to question what I was getting myself into.

At the edge of Hermannsburg the bitumen ran out. Finally I was on the first dirt section of my journey. It galled me that it had taken until here to get off the tar, but unfortunately the death of the laptop had thrown my earlier dirt bashing plans into chaos. However, it sounded like the Mereenie Loop would make up for the lack of dirt roads so far. I'd heard it was a rough trot and the advice was 'take it easy'. The bike, though, appeared not to heed the cautioning. It insisted on cruising at highway speeds across surfaces that would leave the innards of the average family wagon pasted all over the road. I wrestled with the bike, trying to keep the speed down. Admittedly it was a half-hearted attempt, as I loved flicking along the road as much as the bike. All in all, it was a great combination for something to go pear-shaped and as confessed at the start of the book, it duly did when my top luggage box parted company and began its own separate journey.

Other than a lot of 'equipment anxiety', the only real problem caused by the breakaway luggage was how to get going again. One option was to strap the box back on, but that would have meant riding slowly and constantly fretting over the box's potential to go on its own independent sojourn again. I decided that strapping it on was the last resort—the risk of it coming off again was too high and I wasn't confident the equipment would survive another bouncing bomb stunt.

While sucking in flies and thinking 'what now?', my eye roamed across the bike. Bike riders often have obsessive relationships with their steeds, and this is particularly so in my case. This was largely because it was my office, home, partner and packhorse for months on end. When not on-board I ogled it a little more furtively than average riders would their own bikes because in the back of my mind I thought, 'if it shits itself out here, I'm stuffed'. Looking for problems in advance was my way of averting that. My pupils expanded to saucer size when my eyes reached the rear wheel—large chunks were missing out of my new back tyre. I had fitted the tyre in the Alice, and only 200 k's later it looked like it had done a lap of the planet. The road had been much rougher than I appreciated.

> Bike riders often have obsessive relationships with their steeds, and this is particularly so in my case.

Stranded on the side of a road anywhere is never a fun experience, the fun and remoteness being inversely proportional. If the box had broken off somewhere truly remote I would have had only one option: sort the problem out myself. Fortunately the Mereenie Loop is not that remote. Adventurous travellers use it as a back road between Alice Springs and the Rock. The problem is these are generally brain-dead gits in rental vehicles, not au fait with the basic bush courtesy of stopping and checking if someone pulled up on the edge of a road is ok. Half an hour later the first vehicle I saw proved the theory. It thundered past, showering me with rocks and dust, its occupants staring like I was a dangerous part of the fauna that should be avoided.

When my fit of coughing and swearing abated I decided to put my time to use, so I set about organising a replacement for the broken rack. It's amazing what you can do with a satellite phone. A few quick conversations and probably $100 in calls later, I had arranged for a new rack to be chucked on the next available plane; it would be waiting for me at the Rock the following day. An hour later an outback tour truck came to the rescue. Not only did they redeem my faith in people stopping, but they also agreed to cart my luggage to the next roadhouse, which was Kings Canyon.

Pre and post corner speed advice on the Mereenie Loop.

As I rode off, chasing the bus and my luggage, I pondered the notion of how small the planet was becoming. In the middle of nowhere I was able to hop on a phone and sort out the mess. While the technology was something to marvel at, it also made me think that the risk and adventure was being taken out of 'heading outback'.

Fifty k's later I was parked on the roadside, hands shaking and stomach churning, reassessing my opinion on 'risk and adventure'. I'd had a close call while barrelling around a corner—the sun had flared in my eyes and obscured a deep rut, and I very nearly ended up being a mangled heap at the bottom of a valley. It was all well and good to think that gadgets like the satellite phone were amazing, but a fat lot of good it was going to do if I splattered myself across the bottom of a gorge. My tangled mess of flesh and metal would be hidden out of sight and I suspect hungry wedgetails would find me long before anyone else. Sure, I wasn't exactly undertaking an expedition of Burke-and-Wills proportions, but there were still risks. Besides, I reckoned Burke and

Wills actually had a couple of things going for them that I didn't: at least their steeds weren't capable of careering through the bush at quite the same breakneck speeds as mine. And if their mounts did get stroppy and out of hand, at least they could eat the buggers!

Most tourist images show Central Australia as dry and barren, but three years of good rains had spruced things up a bit, and confused a lot of travellers. The pic beneath is not an ad for a dingo disposal operation, just a cheap bit of signage for a backpacker joint.

Permit roulette

Uluru, The Rock, Ayers Rock—everyone has their own preference. I guess I kind of liked Uluru—it seemed a little more substantial. Ayers, after all, never even laid eyes on the thing. The monolith got lumbered with his name because some passing explorers thought it would be a nice thing to do. Beyond the 'Ayers factor', the rock had been called Uluru for countless millenniums and the name actually means something (mind you, 'shady place' is perhaps not the first thing that jumps to mind when you look at it).

I had signed up to embrace the original name and respect local wishes not to climb it. The 'don't climb' thing seemed logical, but logic is a bit like statistics: it can be argued whatever way the wind is blowing. The way I looked at it was you wouldn't go into a church and walk on the altar, so why

Either bored tourists' or locals' imaginative vandalism.

disrespect another fella's beliefs by walking all over their place of spiritual importance. It seemed a matter of both commonsense and respect, but if you don't have either, and you are from the media, it's beaten into you long before you get anywhere near the place.

The terms 'media representative' or 'journalist' have some ugly connotations and personally I prefer the title 'story teller', which conveniently sidesteps a label and also confuses the crap out of people, both of which are desirable outcomes. Unfortunately though, bureaucracies don't usually indulge me in my obsession with labels. In their eyes I was undeniably part of the media. Having 'media' occasionally attached to my name was an ugly reminder that I was not just a traveller; I was being lumped in with the 'muck raking brigade' and this naturally had some flow-ons.

Media representatives who wish to gain access to Uluru–Kata Tjuta (aka The Olgas) National Park must first successfully navigate an 11-page application form. The questions ranged from the obvious, 'What is the purpose of your visit?' to the more 'leave it to the end of the exam' type questions, like 'How will your story contribute to the cultural significance of the park?' Filling out the form does not guarantee access to the park. On my first submission I flunked the 'cultural significance' question and had to do a resit. The applications have to be filled out well in advance of the proposed visit. Television programs wishing to focus on any cultural aspects have to lodge theirs 56 days prior. Even if these 'hoops' are successfully jumped, a TV submission still may luck out as only 3 film crews are allowed access per month.

If the exam is passed, and a permit granted, there are a couple of caveats that go with it. The first is agreeing to a 90-minute session with the rangers on the do's and don'ts of the Park, which includes being driven around and shown

Me and bike in front of Uluṟu—you can't see The Rock because we don't have the rights to show it.

what you can and can't photograph. (Ever wondered why you only ever see photos of one side of the rock? This is because the pretty sunset side of the rock is coincidently the only side of the rock you're allowed to publish images of.) Maybe they feared I could not understand either the myriad of rules or the 'photo prohibited zone maps', but they asked if they could see my material before we published it, just to check I got everything right. Freedom of the press? All editorial integrity seemed to be suddenly going out the window.

Having said all that, I understood—to an extent—where the Parks people were coming from, and it has a lot to do with advertising. Coked up clowns from advertising agencies have for decades been attaching images of this amazing monolith to all sorts of crap: beer, cars, T-shirts—it has even been used as the backdrop in a porn flick. Understandably, the traditional owners have not been overjoyed about all this, viewing it all as belittling the spiritual significance of the place. The exploitation has now been stopped, but I couldn't help feeling that the pendulum had now swung a little too far. There was a raft of detail to be adhered to, and one of my biggest fears was not being able to get my computer to produce the prescribed diacritical marks on Uluṟu and Kata Tjuṯa. Censorship just because I lacked the flaming fonts was too big a humiliation to consider.

Despite the rules, the Parks people could not have been friendlier and more helpful. The compulsory chat and tour basically regurgitated what was in the briefing package. Sadly I suspected they needed to go through the 'one on one' process because program makers/journalists often don't like the details in briefing documents to get in the road of a good story. As friendly as they were, the process and caveats had worn me down. I was thinking, 'I'll just get some snaps for the website and get the hell out of here'.

My stance softened a little when I was finally standing next to the Rock. I had a fresh perspective and realised there was a lot more to the place than just the mandatory 'viewing of the Rock at sunset' depicted in every travel rag. Close up, Uluru took on a new dimension and appearance. Surprisingly the surface was not smooth, but rather a flaky mottled texture, an ancient beast shielded by rocky scales. The haphazard slabs looked as though they could be broken off and nabbed for souvenirs, but fortunately they were hard and therefore safe from memento-grabbing hands. Saying the rock is 'big' is stating the obvious, but when the figure of '9 kilometres' is attached to the phrase of 'walk around the bottom' you begin to develop an understanding. Walking around the bottom not only revealed the shape and contours of Uluru, but it also gave an understanding of the 'shady place' definition. The shadows from the Rock envelop several water holes, cool oases that teemed with finches, vegetation and tourists.

After being in the Park for a day, I realised that all of my heartache over the permits had been for nothing. I wanted my access to the Park sanctioned because I held the vain hope that I might find something unusual—maybe discover some stories and characters that were consistent with my outback mission. Sadly, all I could find was bureaucracy and mass tourism. I saw exactly what the holiday brochures promised—packaged Australia delivered morning, noon and on into the night.

In the evening I wandered into the resort to get a feed. For my money I got some half-decent nosh and a complimentary side serve of the crap marketers love to roll out about the outback. Entertainment was chucked in with the meal, in this case a bloke who played a beer carton and was obsessed with getting all the Japanese tourists to sing 'Waltzing Matilda'.

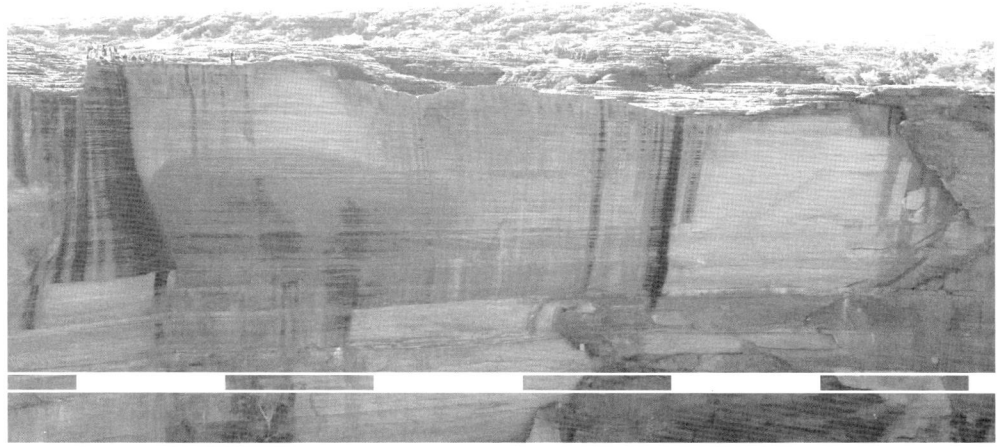

Kings Canyon—complete with tourists, if you can spot them.

'Do you know what Australia's most popular song is?', crackled the speakers.

'Popular' is a fairly loose category, but the Japanese need not have been worried about being put on the spot.

'It's Waltzing Matilda. I want you all to join in with the chorus. It goes like this …' Entertainment Man rolled his fingers across the cardboard and tapped his carton into life. With the rhythm up and running he set about convincing his audience that a suicidal sheep stealer was the greatest musical icon in our short history.

The Japanese looked a little confused by the vocals. I suspected they would have been more comfortable if the lyrics had also been scrolled across a karaoke screen. In hindsight, it was probably best the details weren't on display; a sloppy translation of 'shoving a jumbuck in a tuckerbag' could easily cause offence. Deathly tales of a squatter and swagman were probably just one more bit of confusion for the night. Earlier in the barbecue area the bloke at the 'meat counter' was encouraging them to select bits of the Australian national emblem to chuck on the communal barbie. If roo and emu wasn't to their liking, a host of other bush critters were available to dine on: croc, camel and barra, all the tasty cuts that can be found in plentiful supply at the back of any Australian household fridge.

The 'Australiana' food and entertainment was rolled out each night. There was not much need to vary it as they had a new crowd in nearly every day. The average 'stopover' in the resort is only around 36 hours. Enough time to:
- view at least one sunset,
- do a couple of whiplash tours around the Park,
- neck some cold beer while watching a bloke beat out a tune on Australia's second most popular instrument, the rubber thongs. (FYI: The first most popular instrument was apparently the beer carton. I was interested to see where he ranked the 'bed flute', but sadly it lacked a mention.)

At the end of the day I had got what I deserved by going into a tourist resort. If I didn't agree with what was being served up to the tourists I should bugger off, so that's what I set about doing.

Getting out of the Park required another permit because I was taking the route west, following the Great Central Road, which passed through restricted Aboriginal lands. I was assured the permit process was straightforward. All I had to do was make a quick visit to the tourism office, fill out a form and collect the permit in 24 hours. With the paperwork done, and a day to kill, I rode out to Kata Tjuta to do a bit of rubbernecking and take some more snaps.

Kata Tjuta, besides being a place of great beauty, was also the place where the tar ended and the Great Central Road began. While I was in the neighbourhood I thought I should do a bit of a reccy down the track and get a taste of what I was in for. I stopped and surveyed the track—it looked menacing, with whopping great corrugations jutting up viciously from loamy red sand. A minute later I took the plunge and began shuddering across the top of the corrugations. My innards felt like they were being vitamised, but despite fears I had about fillings being rattled loose, it was manageable. However, when I thought of hauling over 1200 k's of the stuff with the bike fully loaded, the term 'manageable' seemed synonymous with 'pipedream'. I pushed on for around 10 k's to see if the track smoothed out. It did, a bit. This wasn't a huge surprise; the beginning and end of bush tracks are the worst because they get chopped up as vehicles either slow down at the end of a long haul, or speed up to begin one.

Mount Conner. This is on the way to Uluru. It gives rise to a bit of false hope because you see it in the distance and think it is the Rock.

I headed back to the tar and made for Yulara, feeling semi-confident about having the ball skills to tackle the 'Great Central'. On the way back I came across another distance biker parked on the side of the road. Encountering another rider is a bit like bumping into a comrade in the trenches, and the right thing to do is stop for a chat. We discussed the mandatory topic of 'where we had both been'. He told me he had just done the Oodnadatta track and was interested in doing a bit of the Central Road and would like to tag along with me if it was ok. I was chuffed—in 30 000 k's of riding around the country I had never ridden with another rider. A bit of company and support on the start of this epic section was something to jump at. He said the Oodnadatta track had been hell, but he now felt confident about tackling anything else. I agreed and suggested he go and check out the start of the track. We arranged to meet the next day and rode off in separate directions.

Twenty-four hours later we hooked up again. He duly told me I was nuts, there was no way he was going up that bike-wrecking road, then he departed, wishing me luck. It was not the sort of confidence booster I needed, especially bearing in mind his bike was at least 100 kilograms lighter than mine and better suited for tackling the sand. I went to bed hoping that the real issue was that he was just a lousy rider and that I would be ok. The head was telling the story but the heart wasn't listening. Sleep was a long time coming.

The next day my worries about actually having the skills to tackle the road become a minor issue compared to the logistical speed hump put in place by the permit people. There was no problem with getting a permit issued—the difficulty was in the bold print at the bottom of it.

'WARNING: FROM APRIL 4TH UNLEADED FUEL WILL NO LONGER BE AVAILABLE IN TRIBAL LANDS. ADEQUATE FUEL MUST BE CARRIED TO COVER A DISTANCE OF 800 KILOMETRES.'

I was stuffed—the bike was 'unleaded only'. The sale of unleaded fuel had been banned because of petrol sniffing problems; the only other fuels on offer were avgas (aviation fuel) and diesel. There was no way I could do 800 k's on a tank. The best I could hope to get out of a tank on such a sandy surface was about 400 k's, and that was stretching it a long way. To make sure I hadn't misinterpreted I re-read the permit. I hadn't, and I felt like throwing up.

As part of researching the trip I had phoned several months earlier to check on fuel availability and the state of the track. No one had said anything like 'by the way, we are going to suddenly ban the sale of all unleaded.' My head swam. The only other way of getting across to Kalgoorlie was to head back south and cross the Nullarbor. My backers would not be amused by a stuff-up of this proportion, not to mention the fact that I did not really feel like doing a 2000-k detour! In a panic I phoned Justin at BMW (he'd organised the lease of the bike) and asked what would happen if I tried to run the bike on avgas. The response was: 'It will go like stink, but you'll cook the catalytic converter, so don't'. I hung up thinking, 'looks like I'll be buying you a new catalytic converter'.

Before resigning myself to this act of vandalism I thought I should call a few contacts I had on the road up ahead, praying that maybe someone would know of a way of sorting out this mess. Two calls later I was in luck—I found someone who could supply me with fuel part way along. I would still be short about by a 150 k's, but hopefully my reserve stock would cover it.

With it all sorted I went off to fit the replacement luggage rack and begin packing. While tinkering I began wondering why everything had been so bloody difficult lately. Was my attitude all wrong? Maybe I was the source of all the angst. Maybe I was pushing too hard, too fast, rather than going with the flow. When I rode home from England the journey had been full of hassles, but it didn't seem so bad. Maybe that was because I had all the time in the world back then. Also, I wasn't using other people's money and there was no one breathing down my neck looking for results. 'Going with the flow' was not an option on this journey. Then again, maybe this leg of the trip wasn't meant to be. Perhaps I shouldn't haul my inexperienced butt out into the middle of the desert.

Travel guides for the upcoming 'Great Central' had warnings like:

'THIS IS EXTREMELY REMOTE COUNTRY. BE SURE TO HAVE A WELL-PREPARED VEHICLE AND CARRY ENOUGH FUEL, WATER AND FOOD TO COVER ANY EVENTUALITY.'

I was covered in the vehicle department, but as for the rest I had none of it. On the fuel front I had only just enough to make it to my first stop. If I had to double back, or change the plan in any way, I would be stuffed. In regards to water rations, I had about 3 litres, the maximum amount I could fit into my Camelbak waterbag. When it came to food, it was the few meagre bags of lollies I could stuff in my tank bag. Any self-respecting scout troop would have disowned me: I was prepared only for everything to go to plan. This may sound a little on the reckless side, but the simple fact was that I had nowhere to put 'emergency rations'. The bike was bursting at the seams with flaming technical equipment. I guess the upshot was that if I were going to die from starvation, or dehydration, I would at least be able to capture the whole event on tape. Not only that, I could have it all edited and waiting for whoever found my remains. It all sounded totally wacky but I tried to offset that by saying it was also a calculated risk. I had an emergency radio beacon and a satellite phone, so if it all went belly up I would at least be able to contact people and let them know what a dickhead I had been.

> I guess the upshot was that if I were going to die from starvation, or dehydration, I would at least be able to capture the whole event on tape.

Despite the travel warnings, I was also banking on the track not being as remote as, say, the Canning Stock Route. The Great Central did have the occasional vehicle trundling up and down it. All I had to do was make sure I stayed on it and didn't get lost. As I packed the last of my gear I mused on the irony of the situation. I was incredibly fortunate and grateful that the ABC had given me so much rope to play with; if they had only known, though, that I was capable of hanging myself with it countless times over.

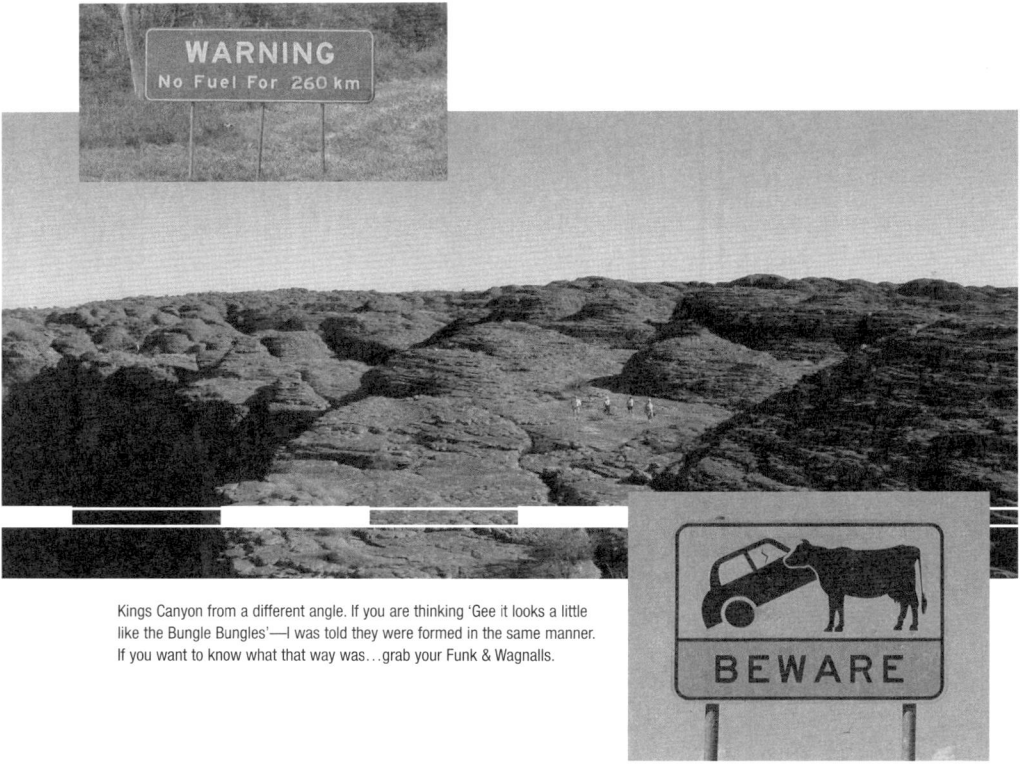

Kings Canyon from a different angle. If you are thinking 'Gee it looks a little like the Bungle Bungles'—I was told they were formed in the same manner. If you want to know what that way was…grab your Funk & Wagnalls.

Photo failures

The Great Central Road is arguably the 'missing link' in the nation's road network. Because it actually exists, I guess it is hard to argue that it's 'missing', but the fact that it's unsealed and wallows through 1200 k's of sand, dust and the occasional set of dunes makes it pretty well non-existent for most normal road transport. If it were sealed, however, it would be a massive short cut for anyone wanting to get from southern WA across to the north-east of the country.

'The Gunbarrel Highway' is also the name a lot of punters like to ascribe to this track. Using the Gunbarrel name instead of the Great Central is a little arse up, as the truth of the matter is there is only a short section of the Great Central in common with the real Gunbarrel. Nonetheless, people like to use the title and who am I to try and wreck their fun. (However, these are also the people

I usually don't accept advice from when it comes to directions.)

At the start of the Great Central there is no sign declaring this is the road you are about to embark upon. This was unfortunate, as the 'suits' back at ABC Mission Central had sent through a request asking, 'Can you take more photos of signposts? People will know where you are when they look up the website'. I seriously doubted many people would've heard of half of the places on my itinerary—I usually hadn't until I rode past them—but who was I to tar others with my own brush of ignorance. I politely pointed out that outback tracks were not renowned for their abundant signage but I would do my best. While the track lacked a Great Central sign, it did however have one for Docker River, the first town along the track. In my new 'a sign is a sign' frame of mind I stopped to take a snap.

Docker River, the first stop on The Great Central Road, a.k.a The Outback Highway. I managed to break a few lights taking a photo of the fact that I was about to start travelling along it…I was looking forward to the damage I could do when I actually got on it.

Photos of signs are, at best, moderately boring. My stunning idea, which would not only liven them up but also prove I was there, was to include the bike in the shots. With my new 'bike in the photo/proof of presence' mind-set I carefully parked under the sign. I took time positioning it, as I was scared it might topple from the weight of the luggage and extra fuel. In the midst of reeling off a few snaps, my side-stand decided to succumb to that Murphy bastard and his poxy law and collapsed. The bike crashed over and slid down a small embankment, stopping at the bottom, partly upturned with its wheels in the air like the legs of a dead animal. Instead of oozing blood the bike was haemorrhaging precious fuel. With fuel involved, super human strength became available. In a single bound I was next to the bike and with vein- popping grunts I wrestled all 350 kilos into the upright position. With the crisis over, my muscles returned to flab and I began gasping for air. My mind wandered past my heaving lungs, over an aching back and on to the bike—it was a second or two behind my eyes, which were already on the scene assessing the damage.

I wasn't even on the flaming track and already I had managed to break off a spotlight and an indicator. I thought, 'If I can do this much damage while stationary on the bitumen, who knows what I'll be capable of when I am actually riding on the dirt.' I pondered over whether this was another omen warning me not to take this track. My gut was in a knot as I began to administer some 'gaffer tape' first aid to the broken lights.

With both pride and bike slightly bruised I checked that everything else was secure and then chucked my leg over the seat. I gazed down the road and looked for a distraction, anything that would let me procrastinate a bit longer and delay getting on the track. There was nothing.

The Great Central begins at a T-Junction, which was inconvenient because I couldn't hit it with a bit of speed, the essential factor for successfully negotiating sand. (The theory is that, like a boat, you need to get up to planing speed and skim across the surface—at this point the bike becomes more manageable. The trick, however, is having the confidence to believe this will work.) Although I had experience in riding trail bikes in sand, never before had I tried to take a bike so monstrous and fully loaded through such conditions—1200 k's of such conditions. I was definitely feeling at the 'chock your jocks' end of the stress meter.

> I thought, 'If I can do this much damage while stationary on the bitumen, who knows what I'll be capable of when I am actually riding on the dirt.'

Unless you've tried to ride either a pushbike or motorbike on sand, your empathy for me trying to negotiate the stuff on a bike the size of a small car is probably going to be a little light-on. Sand does nasty things if not tackled the right way. Essentially it tries to grab the front wheel and turn it in any direction other than the one you're trying to go. While this is unsettling enough in itself, if you throw in whopping corrugations that send jackhammer vibrations through your body, then the boxer shorts are in danger of taking a real caning.

Having run out of negatives to think about, I twisted the throttle and launched onto the track. Corrugations and sand immediately began duelling for the honour of tearing the handlebars from my grip. My teeth gritted and eyes bulged as I wrestled the bike up to battle speed. Once up and planing, the reassuring words of 'you're going to be chewing sand' started popping into my head. My confidence grew a little when I passed the reccy point I had ridden to the day before. I was praying the worst of it was over. Divine intervention, however, was in short supply; 40 k's on and the corrugations were still hammering at the bike. Finally it protested by exploding the protective stone grille off the headlights. I watched it fly forward and then disappear under the front wheel. I braced myself ready for the blow-out; the grille's mounting arms were long and pointy, ideal for tearing into a tyre. Fortunately the pop didn't come, so I set about bringing the show to a stop to retrieve the grille.

> 'Struth, stop your whingeing. Sure it's bloody hard but you've got the job of the century. Hop on the bike and enjoy the bloody ride.'

Walking back to the bike with the mangled grill in hand, I again contemplated the portents that seemed to be gathering. I knew that in the grand scheme of things my journey was not exactly in the same league as Scott of the Antarctic. However, every journey has its hurdles, the size of which is relevant to experience and preparedness, two things I was running a bit low on. Problems, however, and overcoming them, were all part of the psychological game played each day; it's the continually playing it solo that can get a little taxing. It was my second year on the road and with no travel partner to urge me on or tell me to stop being a wuss, putting a positive spin on things didn't always come easily, and even less easily when battling through demanding conditions. The conditions would have been a blast on a trail bike, but no sane person tackled this sort of stuff on a whopping bike that was packed to the gunnels with fragile technical equipment—gear that demanded I constantly hunt for the smoothest part of the track so that it wasn't vibrated to pieces.

As I hunted for somewhere to strap the bent grille the self-admonishment began.

'Struth, stop your whingeing. Sure it's bloody hard but you've got the job of the century. Hop on the bike and enjoy the bloody ride.' Nothing like a bit of Ying to even out some negative Yang. I flung my leg over and wound it up to cruising speed.

The day went on in a pendulum fashion: one second everything was brilliant, the next the whole show was a bee's dick away from going pear-shaped. At the top of one of the more substantial dunes I stopped to take a photo, just so the folks back home could get some idea of what I was wading through. After I had a few shots of the blazing red dune in the can, I put the camera away and fired up the bike. Getting going again is the hard part about riding on sand. I let the clutch out and the back wheel began to dig downward. The bike was trying to move forward, but it was stuck in one spot like a drunk urging his legs to move but nothing happens. Eventually the command to move forward was received at the back wheel and the bike slowly climbed out of the hole. Like alcohol addles the brain, the sand befuddled the bike; it gripped the front wheel and the bike began to stagger from left to right, the zigs and the zags coming in faster succession. Like the truly inebriated, it was eventually overwhelmed by the effort and fell down. As I fished myself out of the sand and crawled back to switch off the engine, I began to realise that I was the source of my own struggles. If I hadn't stopped for the photo I would have

> Like alcohol addles the brain, the sand befuddled the bike; it gripped the front wheel and the bike began to stagger from left to right, the zigs and the zags coming in faster succession.

61

cruised over the dune and kept merrily ploughing along the track. With this incident, plus the earlier sign photo flop, I was beginning to think I should steer clear of photography.

I raised myself from my knees and positioned myself next to the bike, preparing for the clench and lift required to get it back into the upright position. I bent over, grabbed the bars and went for the clean jerk—nothing. I changed my footing, went the jerk and again nothing. Every attempt was thwarted by the sand shifting under my feet. Even if it didn't shift I suspected that the day had started to take its toll and I was now lacking the strength to lift the combined mass of the bike and luggage. When this happens the only option is to unload the bike, put all the stuff on the side of the road and ride a few hundred metres to where the ground is firmer. There was no point in trying to reload while still on the top of the dune as the odds were pretty good that exactly the same thing would happen again. Grudgingly I began to unload—it was the last resort because loading and unloading took time and effort I could well do without expending. A while later I was parked on terra firma, climbing off the bike and beginning the series of hikes back up and down the dune to collect my gear. All this fun while slowly baking in my riding gear on a 35-degree day.

The day progressed and eventually the speedo proclaimed I had survived 200 k's of sand torture. The Great Central seemed disappointed about this and set about cranking it up a few more notches. Throughout the day I had

Desert Parking and 'Yes it is a bushfire' and 'Yes I was stupid enough to ride straight through it'.

seen several bushfires, but none appeared to be in the direction I was travelling—until now. I went into denial mode, telling myself the track would veer either left or right before I got much closer to the smoke that was now dead ahead. Ten minutes, and a few failed prayers later, I was forced to stop—a few hundred metres ahead the road was cut by the fire.

The billowing smoke had turned the blue sky dark and the sun was now a dirty brown dot struggling to be seen through a thick plume. Above the fire front, countless hawks and eagles circled. Occasionally one would break formation and swoop down, picking out an easy lunch from the snakes and lizards trying to escape the flames. It was all very dramatic, but the issue was, 'How the hell was I going to get through?' I could pull over and wait, and hope that it might either change direction or die out, but time was beginning to run short—I did not want to finish the ride to Warakurna in the dark. For a while I sat and studied the fire, eventually concluding that despite the occasional 10-metre flames it really wasn't that big a conflagration. The more I stared, the more I convinced myself that it was nothing to be too phased about. After all, the track itself wasn't actually burning. Speed was the answer, so I decided I would just fang down the road and straight through the middle of it.

Before the ink was dry on the decision I was up to battle speed and on the edge of the fire. It was roughly at this point that I decided the fire was in fact a lot larger than I had originally factored, and also considerably hotter. By this time, however, it was a little late to bale out, basically because it would have taken me about 100 metres to stop, roughly about where the other side of the fire was. This is where one of the main rules of motorcycling comes in: 'When brakes are the enemy, give it the beans.' With smoke pouring into my helmet and my cheeks glowing I charged under the flames arching over the road. With eyes running, and a good round of acrid coughing on the go, I found myself out the other side of the fire, unsinged. It was a good outcome that most normal people would be happy with. I, however, decided I had better stop and get some photos of what I had just barrelled through. For some insane reason I decided the best way to do this was hop off the bike and run back to get some nice close photos. As a safety precaution I thought I'd leave the bike running in case I needed to make a quick getaway. Why I thought I could 'out run' as opposed to 'out ride' it I still do not know, but the decision was made and soon I was about 50 metres from the bike, snapping off a dozen shots.

Docker River, pretty much in the dead centre of Australia. All I was expecting was a petrol pump, so I was a little surprised to find power poles, bitumen roads and phone boxes. I was only surprised because I had broken one of the main rules for travelling: 'never have expectations'.

Fire ferocity depends on fuel load and out in the desert this can be patchy, as bits of bare ground are interspersed with occasional bushes and trees. It seemed I had chosen to approach the fire during a 'bare patch' time. What was a small fire 30 seconds before turned into a roaring blaze with flames leaping out across the road towards me. Instinct took over and I began legging it, but I was weighed down and struggling for speed inside a crash helmet, riding suit and boots. Humphrey Bear would have blitzed me over the 50 and looked far more coordinated in the process. Style aside, I was beating the fire, but by the time I was on the bike the flames and I were again neck and neck. Fortunately the bike was still running; I found first gear, the engine screamed and seconds later I had regained the lead. As the fire shrank in my mirrors the self-admonishments began. 'You idiot, what if the bike had fallen off the stand again, what if you had fallen while running back, what if the bike had caught fire, what if, what if …' Once again I had proven how bloody stupid I could be. Maybe it wasn't such a big risk, maybe it was, but I also knew that my heart was pumping and I was feeling buzzed. With a bit of adrenaline in the system the riding suddenly seemed a lot easier. The bike was no longer taking me along, it was now letting me have some input.

Half an hour later I arrived in Docker River, an isolated Aboriginal community on the NT–WA border. I needed a quick stop to refill my waterbag and also to try and find something a little more substantial than the lollies I had been dining on all day. Turning off the Central and into the town I was surprised to see sealed roads. I thought I had kissed tar goodbye for 1200 k's. Docker River was a long way to haul truckloads of tar, but I guess that just because you live in the middle of nowhere you shouldn't have to put up with dirt roads running past your house. Homes, or should I say the number of them, were also just as surprising. I was expecting something the size of a roadhouse but Docker River was a little town.

In the centre of the town was the community store, which in reality was a besser brick hut. The building was devoid of any windows, or if there ever had been any they had been bricked up. The external walls featured large Aboriginal paintings, the roof featured large and numerous rocks. If 'roof rocking' was an Olympic sport it looked like someone had been training for gold. I climbed off the bike and made for the two-stage entrance. The first stage involved passing through a fence with a spring-loaded gate, one of those childproof arrangements usually found around swimming pools. But it was dogs, not kids, that were the reason for the barrier—it was the last line of defence against the countless canines that roamed the streets. Once past the gate there was a substantial steel roller door that had to be ducked under before entering the shop proper.

The inside of the shop was bleak; scant light struggled to illuminate bare concrete floors and industrial shelving. 'Stalinist' sprang to mind, the difference being that there was a slightly bigger range of stuff to buy here. The shelves were lined with an enormous amount of junk food. If homo sapiens were meant to live off chips, lollies and soft drinks, this place had all the basic food groups covered 100 times over. It was a nutritionist's nightmare. However, there was some salvation to be found in the chest freezers: at one end, piled on top of each other like logs, were enormous unskinned kangaroo tails. Someone had done a bit of 'value adding' by wrapping each tail in a bit of glad wrap. I couldn't help but wonder if meat still packaged in the skin of its original owner was a selling point many city butchers were missing out on. I dragged my attention away from the roo bits and re-focused on my original agenda—scoring some food. For the last 100 k's I had been fantasising about bagging some sandwiches. Sadly they remained a fantasy, as freshly prepared food was not in the store's line of business. I guess I should not have been surprised—I was in the middle of a desert in the middle of Australia and market gardens and fresh fruit are a little light-on in these parts. I resigned myself to this bit of reality, grabbed a few bags of chips, a couple of litres of water and made for the checkout.

Sitting out the front of the store, scoffing my BBQ chips, I was surrounded by hungry dogs, half of which were pregnant with their teats dangling centimetres above ground. The dogs stood between me and the town's public phone, which rang incessantly but was ignored by all who passed. The Telstra logo on the phone seemed to be the only assurance that I was still in same country, everything else was foreign. Of course I, like most Australians, had seen Aboriginal communities, but only as images delivered into my home courtesy of newspapers and television. Like most people, I had never experienced a community like this. I could not help but wonder how, in such a developed nation, we had pockets left so far behind, and where the responsibility lay in fixing it. The questions were ones that had been skirted around for decades. All I knew was that the last time I had seen conditions like this was when I was travelling through developing countries. I chucked my leg over the bike and rode off thinking, 'If this is the Lucky Country then Docker River was dipping out on reaping the harvest'.

Semi-retirement and cross-dressing

'Well, I'm closer to 70 than I am 65. When I get a bit older, I'll go and have a look at the outside, the coast. While I'm young, well young enough, I'll stick to this stuff. This is where Australia is, where the beauty is, this red country.'

The words filtered through Pete's beard and floated up past the rim of his hat. The broad brim darkened a face that that would have been hard to put a number against without Pete volunteering the info. Helping shield his face from further inquiries were a large pair of spectacles, the only betrayal of an armour of swarthy self-reliance. The ensemble was completed by a heavy-duty work shirt, dark blue to hide the dirt, cut-off jeans for shorts and a pair of thongs.

I met Pete at Warakurna, a dot on the map about 350 k's west of Uluru. Warakurna is first and foremost an Aboriginal community. However the small township is hidden behind a hill and well away from the prying eyes of punters who call into the roadhouse. Besides fuel, and one of the cleanest minimarts in Australia, the roadhouse also offered some spots for pitching tents and a couple of rooms for rent.

Pete had been out in the scrub for a few weeks and had popped in to take advantage of one of the key features of the roadhouse—the showers. Most people coming in from a few weeks of bush bashing would be more worried about stocking up on supplies than showers. Pete, however, was sorted in that department; he was self-reliant to the point that the 60% premium on fuel was a source of enter-tainment. The last time he had filled up was back in Adelaide, as his 4WD could do 3000 k's between fuel stops.

Outback Pete, arguably one of the most self-reliant buggers I have ever come across. Packed inside the hand built alloy pod was food for a few weeks, 260 litres of water, over 500 litres of diesel, a bunch of spares and also supposedly a bed.

Pete's Landcruiser was a flat-out desert-buster. Mounted on the back was an alloy pod that not only served as his home, but also stored 560 litres of diesel, 260 litres of water, a huge store of food and 2 of his 4 spare tyres. Somewhere in the middle of it all was a mattress. Pete's rig was a handcrafted bush survival wagon, built to spend its life a long way from tar. The folks in Toorak would have been horrified to see a 4WD used in such a manner. Tar, however, held little interest for Pete; he spent the bulk of his time in the outback, travelling for both pleasure and work.

'I'm as retired as you can get. I can't stop, I'd go mad. You vegetate and you're dead in 2 years', drawled Pete.

In between personal sojourns into the bush, Pete spent his time working for oil companies, guiding crews into remote locations to conduct soil and survey work. It was work he'd been doing since 1969 and he was still in demand, largely because he knew what he was doing, but also because there is little competition.

'It's too wild, too remote, too risky, people don't know the country. You don't have to know the county, but you've got to know how to behave yourself in it.' Pete looked like he knew how to behave himself, but I wasn't convinced about his thongs being the ideal footwear for poking around virgin bush.

I was having trouble reconciling a few things about Pete. On the one hand he certainly looked like he knew how to behave himself in the bush (even if I was a little dubious about the thongs); on the other hand he was doing this at nearly 70 years of age. I was struggling with the combination and thoughts began to creep into my head like, 'Are you ever worried about your personal welfare, like if you fall down?' Unfortunately my thoughts often simultaneously manifest themselves as words, and Pete let me know quickly that he was a long way short of needing a walking frame.

'No. I think if you start worrying about that, then you shouldn't step out past your front door. If you worry about dying or getting hurt you might as well go back home and stay there', he said.

Clearly I had missed the point. Being in the outback was exactly the reason Pete didn't feel, nor look, his age.

'This makes people live longer. Your brain gets out of that rat race, where everyone is trying to eat you, clamber over the top and beat you. Once you get out here, you are your own boss. If you don't want to go any further today, you shut her down, brew some coffee and tell a few more lies', Pete said chuckling.

'So you reckon the outback is the Fountain of Youth?', I queried.

'Definitely. Once you've got to the stage where you've worked all your life, paid the Commonwealth your taxes, go bush. No one can annoy you out here', Pete replied.

The question I was beginning to wonder about was, 'How far into the bush do you have to go to sip at the fountain?' After all, one man's outback is another man's tourist park. Pete appeared to be of the thinking that 'if more than three white fellas have laid eyes on it before, then it's a bloody tourist trap'.

'Most Australians know the names of the Canning Stock Route, the Gunbarrel Highway, but very few of them know the desert itself, inland Australia itself. I don't know how long it took you to get from Ayers Rock to here, but most come across in about a day. Well, all you've seen is a red road, you haven't seen what's off the road. All you've got to do is stop and set up a camp and see what comes in at night. You'd be surprised what lives out here. But people don't see it because they come through too quick', he said.

I was guilty. Pete didn't know it, but I sure as hell did. Although he took his hat off to me for being out here on a bike, at the end of the day I was still sticking to a track that he defined as a highway. Sure, if I had camped I would have probably seen a load of varmints come out of the scrub and mug my rations each night. Granted, they would be amazing to watch. They would also, however, probably be a little light-on in the conversation department and make fairly crap radio. The mandate for my journey was to interview people like Pete, but I wasn't going to find them if I was hiding out in the scrub. Regardless of that conundrum I still felt like a bit of a fake when I met blokes like Pete. I felt like I was doing the outback the easy way, and the only real worthy way was the 200% hardcore way. But this thinking is part of stupid game of one-upmanship among travellers. Remoteness, hardship and glibness populate the language of the travel snob, stuff like:

'Oh, you've missed the best bit. Sure we broke both axles getting in there, but you really haven't experienced it unless you've camped there for a couple of months. Even if we weren't waiting for new axles to be helicoptered in, we would have still stayed that long regardless.'

The 'one uppers' are everywhere and it was no different when I was backpacking—in fact it was probably worse. I would hear lines like, 'I'm on my third passport'. Comments like this were supposed to make you think, 'Wow, you must have been to loads of countries to have filled up three passports'. However, the truth usually was that they were just dopey wankers who had managed to lose a passport in each of the two countries they had visited.

Spot the girl! Claire, on the far left, was part of a trio that were also taking The Great Central. The two Poms and the Irishman had something else in common with me other than the Great Central, they had also done the ride from England to Australia. The ride through Central Australia was the last part of their journey, they also reckoned it was the hardest part by far.

This was not the case with Pete: he seemed too wise for the one-up game. Nonetheless, I still felt a mix of inferiority and senseless guilt, the type that radiates when being spoken to by a police officer.

Pete decided he had hung around in the big smoke for long enough. He bade me farewell, climbed into his rig and made for the scrub. I was left in a cloud of dust, wrestling with myself over whether or not I was a big blouse for sticking to what Pete considered to be a highway.

My internal conflict did not last for long, and the remedy was quite simple—a few tales of someone else's travel woes. Pulling in after Pete were three bikers who felt very differently about travelling through central Australia. Claire, her fiancé Mark and travelling partner Conner had ridden in from the western end of the Great Central. Claire was unusual in that, sadly, there aren't a lot of women bike riders about, especially out tackling that sort of terrain. I was pretty keen to find out how she was faring.

> I was left in a cloud of dust, wrestling with myself over whether or not I was a big blouse for sticking to what Pete considered to be a highway.

'It's been really hard going. Everyone told us it would be easy going, even other bike riders. We set out and hit sand, after sand, after sand. It was pretty deep stuff and I'm not used to that, so it was pretty hard going', she said, sounding exhausted.

Claire, I suspected, was a reasonable judge of what was tough and what wasn't; after all, she had spent the last year riding her bike from England to Australia, and the Great Central was a continuation of the journey. All up, Claire now had 55 000 k's of motorcycling under her belt and was still in one piece, which is a merit badge in itself considering some of the riding conditions in Asia.

Hooking up with another rider usually warrants a beer, or 10. Unfortunately it was only 9 am and not too many pubs oblige at that hour. And pubs were a fantasy here—we were in the heart of 'permit country', which was strictly alcohol free. I guess you could bring beer in, but on top of the cost of the carton you would have to tack on an extra $5000 to pay for the 'no alcohol' fine. Even without beer, bumping into other riders was something to get excited about because they share the same passion and can relate to what you're experiencing. With Claire this was even more so, as we had both done the same epic ride from one side of the world to the other. She had ridden the same roads, seen the same cultures and both of us had survived every road user between Italy and Australia trying to kill us. The roads in Asia are bad, mainly in terms of dealing with lunatic driving antics. But that aside, they are nowhere near as technically difficult to ride as our outback roads. Surprisingly, you can travel pretty much all the way from London to Katmandu on bitumen. There is certainly nowhere along the way that features 1200 k's of sand and dirt. I was interested in how she was dealing with the off-road riding.

'I was exhausted yesterday. I was exhausted halfway through yesterday because I am not used to these conditions. I was emotionally exhausted as well, because I kept thinking "I can't cope with this". But we kept it upright so it was a big sense of achievement. I am supposed to be writing a diary as I go along, but I have to leave it for days because I am so tired. I don't know how you ride each day and the drag out your computer and work each night', Claire said.

I felt like a ghoul, but it was so good to hear these words. She could relate, understand, and was empathising with me, and I was going to milk it. I had my recorder running and her words were going to help me set the record straight with a few nazis back home who reckoned I was off riding around the country having a fat old holiday.

I am such a selfish bastard sometimes. All I could think about was me. And yet in front of me was this poor girl who looked like she was only just hanging on by the tips of her fingers. Psychologically she was just keeping on top of things. Physically she was close to cactus; I suspected the only thing that was keeping her vertical was all the armour in her riding gear.

It soon became clear that Claire was after some reassurance. She wanted to know what it was like on the road ahead. She wanted to know that the worst of it was over. However, Claire had come in from the western end of the Great Central and I was yet to ride the section she had been on, so I was kind of stuffed on the comparison front. I didn't want to bog her down with minute details like 'I'm not yet in a position to judge'. I just wanted her to believe she would be all right and capable of doing it. This required talking the details down with lots of dismissive statements like 'There's a bit of sand, but you'll be right'.

Claire lunged at my words; they seemed to be what she wanted to hear.

'Other people have told us different stories, but you're a rider, we are inclined to believe you', she said, sounding relieved.

I realised immediately that perhaps I had pumped her up a little too far. I started to try to worm my way out of my previous economies with the truth by saying, 'I think it's all relative to your experience and what load you're carrying'.

Even as the words came out of my mouth I knew that Claire was going to hate my guts in 6 hours time when she was struggling down the track. I had overplayed my hand and downplayed the sand aspect way too much by initially telling her 'it's nothing to fret about'. But what could I do? Her eyes were desperate; she was looking for hope. She had obviously heard the same reports as me—that the 300 k's between Docker and the Rock were the worst part of the

Great Central—but there was no point in me reinforcing this. This person needed building up and I would have done exactly the same if she had been a bloke. Claire needed to believe she would get through, I knew this only too well from my own travels. If she didn't believe that, then she was going to make a meal of it. If I did tell her it was a nightmare, what were her options? Turn around and go back?

I wanted to get off this ethical roundabout, so I hung a left by asking her if she thought it was different for women travellers than for men.

'One thing Mark and I love when we get to a camp site is that often we are offered a cold beer, and Conner has never had that when he's travelling on his own. People don't come up to him because he is a guy on his own. A woman is just more approachable, they are not seen so much as a threat. People offer you more hospitality and generosity.'

'Well I guess I am in the same boat as Conner. Perhaps I need to change my circumstances and consider cross-dressing when I'm on the bike?' I suggested.

'Yeah, it's a good plan', Claire said laughing.

With that I switched off the microphone and let her get on her way. Before letting her roar off I gave her my details, hoping that she might get in contact and let me know how she had fared on the last bit of the Great Central. Sadly I never heard. I suspect she may have binned my card in disgust after dealing with all the sand that I had promised would not be so bad.

10

Urban living in the dead heart

Chatting to people, and generally sniffing around, was pretty much the way I gathered most yarns as I travelled around the country. Occasionally, though, I had to get a little more structured about the whole scenario and plan ahead in minute detail. 'Occasionally' was a synonym for 'Television'.

Television is ridiculously labour intensive. If there are roughly 7 dog years to every human year, then in the world of making television there's approximately 1 human year required for every hour of TV made. I had ridden to Warakurna with the express purpose of involving myself in some accelerated ageing:

I was to do a little yarn for a kids' TV program called 'Behind the News'. The story was about the meteorologists at Giles Weather Station, which was 1 k up the road from Warakurna roadhouse. The story was going to be a bit of a snapshot about why these blokes lived in the outback collecting info on the weather. The topic may not sound all that exciting, but for the program's 1.3 million school kids who mostly live in cities it was a great opportunity to show them a bit of the outback and what it was like living there. The intention was good; the only hiccup was in the execution.

> It all seemed good at the time, but when I was standing behind the camera trying to film a weather balloon as it accelerated out of frame and into the stratosphere, I was thinking a bit more experience might have been handy.

Doing a TV story had been entirely consistent with my ongoing game plan of being 'not prepared and under experienced'. Never before had I shot a TV story, nor had I scripted one, nor edited one. These facts, however, I thought would be an advantage—at least I wouldn't know if I was stuffing up or not. To be perfectly honest I wasn't a total TV virgin: firstly, I had spent half a life watching the stuff (in fact, I think this makes most people way over-qualified for making it); secondly, I'd had a couple of brief brushes with TV when I was used as a talking head to present a couple yarns, but on those occasions the behind-the-scenes strings had been pulled by people far wiser than I.

As I rode up to the weather station I contemplated how I managed to get myself into this mess. As usual, it all came back to the budget and the promises I had made while trying to get the cash together. The 'backers' wanted a larger profile for the journey, so it was decided that tapping into a TV audience would do this and also attract a lot more people to the website in the process. I said, 'No wukkas, I can bag a few yarns'. With the emphasis on 'few', and also omitting the minor detail of never having produced a TV yarn before, I grabbed the money and bolted.

It all seemed good at the time, but when I was standing behind the camera trying to film a weather balloon as it accelerated out of frame and into the stratosphere, I was thinking a bit more experience might have been handy. Balloon launches are one-hit wonders—you screw up the shot then you get to wait around a day for the next launch. I got to be 'carry-over champion' and had to come back the next day. I cursed my stuff-up as I trudged back to the bike to get more equipment.

Each night I would unpack all this and set up my portable production suite. Supposedly it was so I could get some work done, but the truth was I just loved packing it all away the next morning and loading it back on the bike.

Despite cocking up the 'action sequence', I bumbled my way through the rest of the yarn by:

1 Pointing the camera at a meteorologist and harassing him about why he was stationed in the outback and what he did while he was out here,
2 Recording a few shots of the meteorologist walking around in a lab coat (the fact that they never normally wore coats was not something I was going to let get in the road of shooting some meaningful pictures),
3 Filming shots of computers and the meteorologist (still in lab coat) playing with them, and
4 Recording plenty of outside shots to set the scene and detail the infrastructure required to live out there.

By the time I did all this, plus bolted the camera to the bike for the 'talking while riding' bits, the whole day and several tapes had been chewed up.

The filming was in fact the fun part; the nightmare began afterwards when:

1 The video material was transferred into the computer (not a quick process when dealing with $2\,^1/_2$ hours of tape).
2 The plethora of footage was edited into a rough story—this was vital as it showed where the stuff-ups were. (The stuff-ups were easy to find: it was the good bits that were in short supply. Searching for quality shots proved to be a bit of a bugger, not only because of their rarity but also because I was self-taught with the editing software and had no idea what I was doing.)
3 To fill missing gaps in the story I drew up a list of what needed to be re-shot (surprisingly this was longer than the original shot list).

4 Began dribbling into the keyboard at around 2 am when I fell asleep on top of it.
5 Set alarm to rise at 7 am so I could re-shoot balloon launch at 7.30 am.

I arrived back at the weather station not long after sunrise and quickly set about my first task, which was convincing everyone to wear the same clothes (while also trying not to bore them with too many details about 'continuity'). They looked a little bemused over my return and the amount of time this videoing routine was taking, which was kind of understandable considering the finished story would bounce around the airwaves for about as long as it would take to boil an egg. As I wandered around the facility for the second time it became more apparent that life on the weather station existed as a microcosm—it was a self-contained oasis in the middle of a desert. The staff might as well have been living in Antarctica because they really had no need to venture off the base—all their food and supplies were delivered to them. When it came to maintenance, the Bureau supplied a full-time maintenance bloke to look after everything from bores to generators to blocked toilets. The maintenance man had all his materials and equipment freighted in too, which was just as well, as the nearest hardware store was an 800 k drive. The nearest cinema was also a similar distance, so the 4 staff members had all their entertainment needs met in-house with a satellite TV, a pool table and a small library of videos and books. Because everything was laid on, there really was no reason to leave the station, which was exactly what the Bureau wanted. And the reason for this was all to do with safety on outback roads.

Parked under the carport was a swank new 4WD. Its mission in life, however, was restricted to driving a few k's every second day to check on the bore. Off-station expeditions were forbidden, so doing the 700 k round trip to Yulara for an overnight bender was out of the question. The Bureau was not going to let staff members risk their necks, and the truck, out on the Great Central. (Mind you, I reckoned regular beer runs should have been station policy, as being marooned together for long periods may have been more of a health threat than the Great Central.)

The isolation and the self-supporting infrastructure were contrasted by the ordinariness of the main dwelling—essentially it was a Joe Average suburban structure. Outside there was the minor luxury of a small pool, but inside the only noticeable difference was a large(ish) communal living area and kitchen.

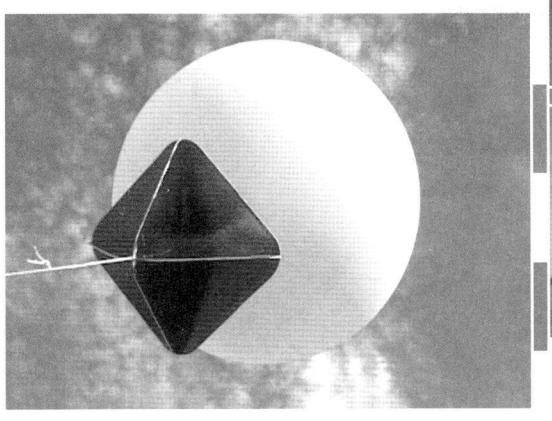

Every single day of the year a meteorologist at Giles weather station wanders out and sets one of these free. Arguable not the greatest spectacle on earth but it's worth a gander if you are passing by.

Admittedly it lacked the personal touches of a home—there were no 'fourth and final reminder notices' stuck to the fridge or disassembled bicycles on the back porch. But despite this it was still closer to being a home than the impersonal transient dwellings found in other remote places like mining camps.

I continued filming, and while my hands wrestled with the camera my mind wrangled with why I had expected the dwelling to be harsh, like the surrounding environment. It was a ludicrous expectation and a total contradiction of the reason I was out here: I wanted to find anything but cowboys and yet I was disappointed that everyone wasn't sleeping in swags. Just because it was incredibly remote didn't mean they should have to live in the same conditions as explorers like Len Beadell and his crew did.

For a long time the Bureau staff did rough it in portable tin huts; the original accommodation block stood at the back of the house as testament. The meteorologists, though, had enough to contend

> It was a ludicrous expectation and a total contradiction of the reason I was out here: I wanted to find anything but cowboys and yet I was disappointed that everyone wasn't sleeping in swags.

Stills from the TV story I shot at Giles weather station. The story was for school kids, so I figured the bigger dill I was in the story the more attention they would pay.

with. They worked shifts of 6 days a week in an isolated location for months on end, so why should sub-standard accommodation be chucked in as part of the equation as well? Yet I also had this strange view that because they generated their own power, pumped their water etc., they would live far more frugally. It was a strange thought, especially considering that the running costs were all picked up by the Bureau. And on top of this, all the metrologists were, like me, from the city. We were all part of a group that on the average day back home was totally removed from the resources that provided our standard of living. Perhaps this answers why Australia, the urban nation, is so good at consuming resources. I was once told that if the entire global population wished to enjoy the same standard of living as the average Australian it would take a bunch of 'planet Earths' to supply all the required resources! Australia has the badge for 'highest per capita greenhouse emissions in the world', so I guess we must know a thing or two about consumption.

'But I'm all right Jack, I recycle my cans', I thought as I packed the camera away. Sure, I might have been blasting around the bush chewing up thousands of litres of fossil fuel, but imagine how much more I would be chewing through if I were in a 4WD! 'Yep, I'm guiltless and I am out there totally in touch with the elements,' I thought as I tightened up the velcro on my riding suit, ensuring that the wind, heat and dust were all locked out. Yep, I was in touch.

POSTSCRIPT It seemed that lack of experience paid off with the TV story. After many days of sweating over the laptop I posted the yarn back to Mission Central. It was a finished item, complete with voice-overs and music. I fully expected it to be re-edited, re-formatted and so on, as no television content goes to air without 10 layers of sub-editing. Oh, me of little faith—they paid me the compliment of putting the item directly to air from the tape I sent back. Got to be happy about that for a first stab.

Passing aliens

I wanted to spend another day at Warakurna but I simply did not have the time. The schedule, or more to the point the budget, dictated that I keep moving. Despite being shagged out of my scone from the previous few days of TV mayhem, I left Warakurna feeling upbeat. I had met a bunch of great people and bagged a wad of material: a TV yarn, a radio story and a bunch of snaps for the website. A reasonable haul for what was essentially a 36-hour stopover. Best of all, I was on the road by around 10, giving me loads of time to cover the 250 k's to Warburton.

As usual the reports about the road conditions between the two places varied somewhat, ranging from a bloke from the community pissing himself laughing and saying, 'You're going where? That road's fucked mate', to

Lush Central Australia (if only you could see it in colour) and one of the many creative ways they like to park cars out on the Great Central.

the fella running the roadhouse who reckoned, 'Compared to where you've come through, it'll be a snap'.

Taking advice on road conditions was always a bit of a lottery; one person's track is another's highway. On this day, either I had graduated to the 'highway camp' or the track was smoothing out. There were still patches of sand but the occasional sets of low dunes had disappeared and the sand was now hidden in the bottom of deep wheel ruts that had been carved into the road and often stretched for many k's. Riding near the ruts was a white-knuckle experience—if I suddenly dropped into one it would have removed my foot-pegs and sent me cartwheeling down the road. The ruts appeared randomly and I viewed them as the bush equivalent of speed humps, ensuring clowns like me stayed on their toes and kept their speed down.

The landscape continued to look lush, thanks to a few years of good rain. There was an abundance of young eucalypts and acacias dotted across vast carpets of golden grass. The yellows of the grass and the reds of the earth were rich and deep in the morning sun. It was enough to make me want stop and spend a few minutes to soak up the spectrum and contemplate my belly fluff.

I worked down through the gears and pulled over under a tree, eager to hop off and park my butt on the ground so I could stare at this beautiful landscape. Millions of ants, however, had other plans. A miniature army had secured all positions in the shade and they seemed to be telling me that dismounting and straying over the verge was contrary to the rules of transit—my alien status was being reaffirmed. I decided I would vent the bladder while continuing to scope for suitable ground. My tight schedule was making me a time-management fascist—even in this brief interlude I was still trying to maximise every moment.

As to be expected in a moment where a little privacy goes a long way, I heard the sound of an approaching vehicle (Murphy's law is definitely universal—I hadn't seen a car all morning). Interrupting 'the business' was bad enough, but more of an issue was interrupting the quietness that had quickly enveloped me, a silence I was revelling in. Five minutes passed with no sign of a car, but I could still hear the low rumble of a distant engine, an engine that sounded as though it was still in the same spot. My 'approaching vehicle' theory evolved to 'a grader working further up the track'. It seemed a good theory but its downfall was that every time I moved my head, the grader moved with it. If I had spent less time trying to be a flaming bush tracker I would have realised that it was my ears ringing from the morning's ride. Even in stillness I was still shattering the peace. It seemed like suitable punishment for my endless roaring through the landscape. The ants and the ringing took the gloss off the vista, so I saddled up and rode off.

> Not knowing where the camels were at in the lust cycle, I decided that pulling up short of them might be a good plan.

An hour later my presence was again being challenged, this time by a herd of camels blocking the road. I knew very little about camels, other than to avoid the buggers during rutting season. I'd been told the males can get quite stroppy and it pays to steer clear of the sods. Unfortunately, the detail I hadn't been told was when exactly mating season was—for all I knew this was the peak hour in the camel lust cycle. If the bull did decide to get the strop on, I needed to keep in mind that I wasn't safely tucked away inside a car (there are good reasons why safari zoos don't conduct tours on motorcycles). I didn't think mauling was likely, but charging, biting and kicking were possibilities to be factored in.

Not knowing where the camels were at in the lust cycle, I decided that pulling up short of them might be a good plan. I hoped that if I waited around for a bit the dopey beggars would eventually wander off the track, but not before I was able to take a few photos. I need not have rushed for the camera; the camels stood their ground and expressed their indifference to my arrival by pointing their backsides towards my lens. After getting half a dozen shots of 'camel dates', I decided that:
- the camels were not going to move, and
- shots of their bums were pretty uninspiring.

With the camels being in 'ignore mode' it was easy to put my 'rutting anxiety' to one side. I decided that the best way to clear the track, and improve my photos at the same time, was to stir the beggars up a bit (I've advanced a lot since I was a kid poking sticks into ant nests to see what the response would be).

More parking action.

Chucking rocks was an option, but I suspected a stone in the head might push them over the edge on the grumpy stakes. The second option was to get on the horn and make some noise—I was working on the fright and flight principle.

The first horn blast scored neither, just some annoyed-looking gazes from the few camels that bothered to turn their heads around. Engine revving and flat-out horn blaring was the next 'track clearing' alternative. They took the flight option, I suspect more out of just trying to get away from the racket than being scared. After all, there were 8 of them and only 1 of me. Their departure enabled me to get a few more photos of their rumps, this time in disappearing mode. As they loped up the track I couldn't help but wonder if they were messing with me; there were millions of square k's of scrub to vanish into but this lot decided to travel straight down the road that I wanted to go along. A few hundred metres of chasing them with the horn blaring was enough for them to decide to leave the comforts of the road and go bush. I caught one last fleeting glimpse of their backsides as they slowly merged with the bushes, and then turned my attention back to the road and continued on towards Warburton.

At the outskirts of most towns around Australia travellers usually pass a sign that welcomes them to the region. Often the sign will also give a reason for the town's being, something like 'Welcome to Giblet, Origami Capital of Sunraysia' type things. Warburton was up for none of this. A large yellow sign erected at the start of the road into town heralded 'Warburton Community, Private Property, No entry without permission'. As far as 'welcome to' signs go, this ranked more towards the 'sod off' end of the scale, which was exactly what the community wanted. The Warburton lot had no time for errant tourists trundling through their streets, gawking and taking wads of pictures. Anyone is going to get a little rankled over their house being viewed as a tourist attraction.

Welcome to Warburton.

Long ago the folks in Beverly Hills built some of the tallest fences in the world to deal with this problem. Fences, though, are a little contrary to the way this mob live, so the easiest solution was to place a big sign on the outskirts telling people to mind their own business. I suppose a cynic could say 'Shouldn't they be proud and want to show off their community?' The Warburton mob were of a similar opinion, the difference being that they wanted a dedicated space to do it in, so they built a whopping great cultural centre next to the town bypass. The centre meant travellers didn't have to take a detour through the town and annoy the locals. All they had to do was pull over to the side of the Great Central and wander in to find out about the community, culture and art.

I eyed the centre from the back of my bike and decided that I would pop in there the following day. The ride had taken its toll—all I wanted to do was get into the tourist camp and sort myself out with a shower, a bed and a bit of kip.

FERAL CAMELS

The western half of Australia is well and truly camel country. There is no denying that a herd of bloody great big camels makes a speccy sight. The downside, though, is that they are nothing more than giant-size rabbits wreaking havoc on the bush. Camels are about as natural to the Australian landscape as motorcycles, but the problem is that the camels are breeding at a far more prolific rate. Between 600 000 to 750 000 feral camels are estimated to be roaming around the bush. In relation to rabbit numbers that's small bickies, but your average rabbit doesn't trample down fences, foul water holes and beat up on sheep and cattle (apparently they have a reputation for charging livestock, pushing them away from areas of feed and water).

Of all the waves of pests that have plagued this country, these pests are only just starting to realise their true potential. One estimate is that the population will peak at 60 million, outnumbering roos by 100 to 1 in some sections of the bush. The only worthy predators camels face are trucks and trains, and the trucks are coming off poorly in most camel clashes. (This is because most wise truckies will do anything to avoid hitting a camel—being on the tall side the buggers have an unfortunate tendency of coming through the windscreen.) Usually the only solution is 'the swerve', but road trains aren't renowned for their agility at 100 kph. The truck usually rolls and the dopey camel trots off, unscathed and wondering what all the screeching and crashing was about.

12
Dressed to build

Travellers are faced with what some consider a blessing and others a curse, and it all has to do with wardrobe space. The average traveller has little room to carry a huge array of clothing, and really what is the point if you are on the move and with different people every day. For me, the conundrum of choice never featured in any 'getting ready' morning angst—I had one riding suit and that was it. If I wanted more I would have not only been vain but stupid as well, as riding suits cost a bomb and they are bulky armour-plated items that would require a trailer to carry a spare in. One suit is enough, and it has to be a good one. Firstly, a good suit maximises your chances of walking away from a stack, and secondly, it helps shield you from weather, dust and grime. The problem with living out of one suit for months on end is it gets a little manky. Dirt,

Just to prove I was there.

insects and filth grind into the suit, day upon day, layer upon layer. Even the most loving mother would be hard-pressed to give a welcome home embrace to a long lost son encrusted like this.

Fortunately though, the outback is a forgiving place when it comes to dress sense. This probably comes from people's attitude that there are more important things to worry about. They probably also know that togging up just increases the chance of having to climb under a car and fix something. For me, permanently clad in riding gear, the dress standards were perfect; I could walk anywhere, looking like a filthy storm trooper, and the only comment I would occasionally get was, 'Aren't you hot in all that?'

> The fundamental element of bush dress sense is 'keep it minimal'.

The fundamental element of bush dress sense is 'keep it minimal'. It's all about staying cool. However, as I took stock of a woman in a denim skirt that finished two threads below the belt loops, I thought this was perhaps taking it to extremes. Above the skirt's waistband, way north of the belly button, was the start of a white halter-top, the spotless cotton calling it quits well short of the neckline. The outfit was finished off with a coy acknowledgment to the outback—a broad-brimmed hat at one end and RM boots at the other. All in all, plenty of exposed skin for things to bite, for the sun to burn and for eyes to roam.

The flesh belonged to Deborah, the camp cook for a construction team. The team was returning from building houses for an Aboriginal community at Nyapari, a dot on the Old Gunbarrel Highway south of Uluru, and a long 4WD trip from anywhere.

'Being the only girl on the building site is alright because I like to act like a bloke', Deborah said laughing.

'I'm not a girly girl. I'm conditioned to it, so I don't feel out, I don't feel like a female.'

Either Deborah had failed to look in the mirror in the previous decade or she was taking dry humour to new drought-stricken levels. Either way, I was having trouble buying it.

'But the way you dress and look, you don't look like one of the boys. Surely the other blokes that you work with look at you?' I blurted out.

'Oh yeah, I'm bad for that, I make sure I tease 'em', she said, bursting out laughing. 'You don't think I walk around in short skirts for nothing?'

A tease is one thing, but I imagined the builders she was working with might have viewed it as 'cruel and inhumane'. Deborah was the only female on a construction team that was isolated in the bush for months on end. It was a harsh working environment and the blokes in the crew had many frustrations to deal with. Separation from their partners was just another one in a long list of frustrations, but I suspect it was pushed towards the top of the list, thanks to Deborah's efforts.

'Doesn't that cause hassles?' I asked her.

'No, no, not really, it's all just good fun. It's all just for a laugh', she said.

Deborah sounded sincere, however I wasn't convinced the blokes would view it the same way after a couple of months of isolation. In truth, Deborah's dress sense wouldn't have even been noticed in the city, but in the Warburton camping ground, a destination only a blind person could call sexy, she stood out like a skinhead at a Tupperware party. I wanted to know how she dealt with the frustrations in the camp. Were there rules laid down by the boss—'don't screw the crew' type rules?

'It's up to me, whatever I do. You just have to be careful who you choose. Yeah, it's up to me', Deborah said simply.

'It doesn't cause tensions?' I questioned.

'Nah, nah, although by the end of the stint, the way I act, it's a bit rough on the blokes. They all want to get back to their missus because here's me flaunting around', she said, breaking into laughter again.

Call her briefly dressed? Yes. Call her candid? Bloody oath—I think. I guess what was disarming was that Deborah seemed to have a male's attitude towards sex—'all talk'. I guess it made sense, as she openly admitted she was a bit blokey, but that was hardly surprising considering she had spent the last 4 years working in the outback with teams of blokes.

Deborah may have described herself as a cook but she didn't just confine herself to the camp kitchen. When she wasn't slaving over a stove she was out working on the construction site, labouring, digging, hammering and so on. Working on the site and roughing it in the same conditions as the rest of team helped her to achieve something else: it earned her the respect of the blokes she was working with. She also said she liked labouring because 'I like to keep my body happening'. I suspected, however, that 'respect' and 'body tone' were not the only reasons she was out there.

'I love being out in the bush. I can't seem to function in a city. I can't manage myself with money, cope with deadlines and all that. I find the city life just petty. Everyone has a little problem and it all just gets blown out.' Deborah's discomfort with the big smoke was reflected in her voice—her speech became staccato and her words quavered, agitation betrayed by a faltering larynx.

Working out in the bush and dealing with harsh conditions was more than just a way of avoiding the city. For Deborah, being out there was also about empowerment.

'I don't want to be looked upon as useless. I look at my friends and they can't even lift a hammer. They wouldn't even know how to use a cordless drill—they'd ask a bloke. I like to be able to do it myself and not rely on a bloke', she said with conviction.

Deborah's self-reliance wasn't limited to construction. Earlier in the day I watched her poking around under the bonnet of a 4WD. The vehicle was desperately in need of some attention: the front of it had been all smashed up. The windscreen was cracked, the headlights were shattered, and there were more stone chips on the bonnet than original paint—it looked as though someone had spent a day throwing rocks at it. In fact rocks had been hurled at it for a day after the ute

L to R: This is what happens to cars that fail trying to traverse the Great Central, they get butchered and used as paddock bashers

Bush Hiltons, many a luxurious night have I spent in dongers like these.

had broken down on the Gunbarrel Highway and had to be towed back. Deborah and her boss endured a torturous day being towed 500 k's back to Warburton—hour after hour they bounced around inside the ute, continually bracing the windscreen against huge rocks kicked up by the tow vehicle. Not exactly the normal day at the office, but I suspected that there was no such thing as a normal day for her out here. I asked if her time out here had helped her build an understanding of the bush.

'I haven't learnt so much about the bush, except that I know that I am not scared of it', she answered.

'Do you think a lot of Australians are?' I asked.

'Oh, they'd shit themselves. They just have no idea about the conditions that are out there', she said, shaking her head.

With that Deborah said goodbye and walked out to the main road. She had a few hours of sunning to get in while waiting to hail down their company truck that was due through later in the day. With the ute being cactus, and a long wait for spares, the plan was to winch it onto the back of the truck and head for Perth.

I would be following the same route, but there was a matter of a kiln to attend to first.

13

Glass art dreaming

'Warburton generates its own power, doesn't it?' I asked.

'Yep', replied Albie, looking up at me from under a mop of long curly black hair.

'This kiln is massive. All the lights in town must dim when you turn it on!' I quipped.

Albie laughed. The question may have come out sideways but I was halfway close to the truth.

'Just about. We actually have to fire it up at night time when everyone is sleeping,' she said.

A kiln, big enough to cook a king-size bed in, was not on the list of things I expected to see while bashing around the bush. Then again, neither was

Aboriginal art made out of glass particularly high on the list; if it had been, I guess I would have expected to see a kiln.

The kiln was part of the Warburton Arts Project, a nice simple title but a little short on zest and explanation. Behind the label was a plan to help preserve the local Aboriginal culture through art. The part I was having trouble with was that for about 60 000 years kilns had not been a part of Aboriginal life. With this being the case I wondered how 'glasswork' fitted in with traditional indigenous art.

Outback marketing propaganda occasionally features Aboriginal painters and paintings. I am pretty confident though there aren't too many tourist brochures featuring Aboriginal glass artists. The lady standing in front of the filthy great ceramic kiln is Dorothy. She was about to bung a bunch of glass in, flick the switch on and place a serious tax on the Warburton generator.

Albie was the coordinator of the project and quickly set about broadening my mind.

'I guess acrylic painting is not a traditional medium either', she reminded me. 'However, it's all stuff that comes from body painting and rock art. I guess we've now got contemporary ways of being able to express those things. When it was introduced, people were very excited. It seemed a natural progression to move from the thick lines painted on canvas on to another medium.'

I had met Albie earlier in the day at her office, a side room off an exceptional looking building known as the Tjulyuru Cultural and Civic Centre. I say 'exceptional' because an architect had carefully designed it. This might not seem all that exciting, but in the middle of the outback it is. To get materials and labour out there requires NASA-type budgets—no one can afford to jack up costs further by involving an architect and a slick design. A concept like an 'inverse veranda'—one that points up as opposed to down—would never get off the drawing board for the average bush building. However, when it came to the Cultural Centre they liked the idea so much they used it on both the buildings that made up the complex. The inside of the building continued in the same unique fashion: fittings were hewn from local wood and the floors

Warburton local Lalla patiently showing the nosey bloke on the other side of the camera how to make glass art.

featured river pebbles encased in polished concrete. The purpose of the building, other than housing Albie's office, was to showcase the culture and the art of the Wongi (Ngaanyatjarra Aboriginal) people —or for people like me who can't get their mouths around 'Ngaanyatjarra', the fall back position is 'the mob that live round Warburton way'.

When I arrived at Albie's office I was given a tour of the gallery, a cuppa and then packed into a car. Albie decided the best way for me to get a handle on what was going on was to check out the glassworks and meet some of the artists who worked there. The glassworks were in the heart of the community. To get there was short blissful drive along a bitumen road—Warburton, like Docker River, also had tarred oasis status. After the Cultural Centre, the glassworks were a stark contrast: they were housed in a large corrugated iron shed, dark inside and oppressively hot during summer. It didn't strike me as the dream environment to inspire great art. Leaning against the shed's inside walls were stacks of glass, the panes ranging from door size down to small platters. The range of choice was further multiplied by a myriad of textures and thicknesses. (I suspected the most difficult part of the creative process was deciding which piece to work with.) Between the glass stacks were the occasional outcrops of shelving, each crammed with white fibrous mats, works in progress and more raw glass. Sitting back from the shelves and stacks were several work tables, each protected with thick green pads and all heavily scored with deep knife marks. All the elements fitted together as an ad hoc production line; infrastructure to feed the kiln in the corner.

Standing in the midst of the assembly line was Lalla, Warburton local and glass artist. Lalla was probably not the tallest woman in town, but what she lacked in stature I suspect she made up for with clout. Her T-shirt proclaimed 'Women's Council' and featured a motif of an Aboriginal woman surrounded by the names of the mobs that made up the council. Council member or not, she had her own independent authority; inquisitive heads that kept popping around the door to see what was happening were quickly sent on their way by a few barked words. Lalla wasn't up for interruptions, as she was keen to tell me about the glassworks.

'I did this for many years. I don't know how many years', she looked questioningly to Albie.

'Six, seven years', coached Albie.

'Six, seven years', Lalla echoed back, 'and I enjoy doing this one.' Lalla's English featured the occasional dropped word, the occasional additional one. As English is often a second language in communities like Warburton this was not all that surprising, and kind of made sense—after all, if you live in your own community why not speak your own tongue?

Lalla pointed to her latest work, which was destined for the oven.

'It's a story about the bird gathering leaves and branches together, like…um, like them big birds.' Again Albie came to Lalla's aid with 'bowerbird'.

'Yeah, bowerbirds', Lalla said.

Lalla's bowerbird work had originally started life as a painting, a medium in which Lalla has some significant works.

'I've got a big painting, it's at the parliament house in Perth', Lalla said proudly, and a smile snuck across her face. Smiles stood out on Lalla, drawing attention to her remarkable complexion. She was a middle-aged woman who had spent her life in desert country, so having flawless smooth skin was quite an amazing feat considering the abuse meted out by the central Australian sun.

Albie and Lalla told me that a painting was where most glass pieces began. From there they are traced out and copied onto thick non-flammable

> As English is often a second language in communities like Warburton this was not all that surprising, and kind of made sense—after all, if you live in your own community why not speak your own tongue?

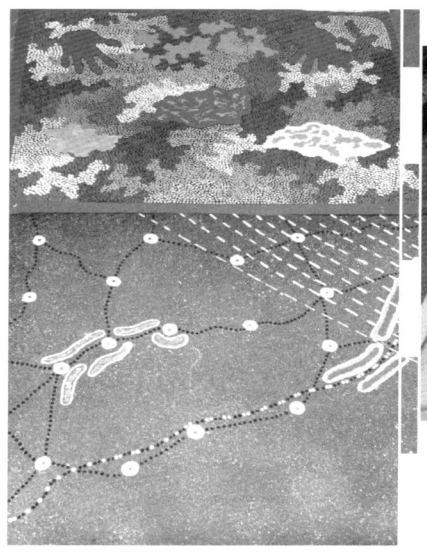

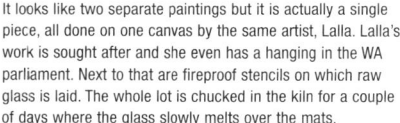
It looks like two separate paintings but it is actually a single piece, all done on one canvas by the same artist, Lalla. Lalla's work is sought after and she even has a hanging in the WA parliament. Next to that are fireproof stencils on which raw glass is laid. The whole lot is chucked in the kiln for a couple of days where the glass slowly melts over the mats.

fibre mats. The images are then cut out of the mat and arranged on a heat-resistant surface. Then comes the consuming task of shifting through the stacks to find a piece of glass. When the agony of selection is over it's then laid across the shapes, shoved inside the kiln and left to bake. The 'baking' process is slow, as the glass has to soften and mould itself over the shapes. Forty-eight hours later the piece is pulled from the oven and the former flat surface of the glass has been impregnated with the shapes of the design.

Albie and her fellow artists produced pieces that ranged from large wall hangings down to small plates you could bung some bickies and dip on. Though bunging a bunch of Jatz on one of these might require a second thought, as the pieces weren't cheap. Despite the price, Lalla, like most artists, rarely recouped the cost of the labour she put in, and the graft was especially onerous considering it's carried out in a tin shed on days where the outside temperature was regularly over 40 degrees. When you take a few of these things into account, buyers not only get value for money on the labour front, but they also get a one-off piece that usually tells a story of the artist's dreaming.

Ensuring survival of these stories was what the project was about, or in Albie's words: 'It's a cultural maintenance project. It's designed to help maintain and strengthen traditional Ngaanyatjarra expression.'

This may sound a little 'departmental', but at the core it's about trying to capture and build on what remains of their culture, because (to put it bluntly) it's dying out. Telling the stories in contemporary mediums makes it more interesting and relevant to kids growing up in Warburton. However, to achieve their aim, one minor snag needed to be overcome: glasswork, pottery and painting were considered a little too 'crafty' by the young blokes, so Albie and her team set about getting around this by establishing a music arm to the project. They provided instruments, tutoring, and a studio in which to thrash, record and mix. The upshot has been that younger generations have found out more about their history and are keeping it alive in the process.

My shoddy attempt to show what the glass looks like two days later. Continuing in my bid for 'crap photographer of the year award' is a wonderfully natural shot of the head of the local pottery works, Lydia, who, with some help, was showing off some work that was about to be shipped to Sydney for an exhibition.

Besides the cultural preservation aspect, the mediums have also tapped into new art markets. Warburton glass is now much sought after and fetches a good dollar. The cash factor is something that can't be underrated, as employment opportunities here are few. Becoming a 'shire man' and working on the road crews is one of the very few local careers.

99

The project wasn't just contained within the studios—regularly the elders would take young groups for week-long trips out in the scrub and teach them traditional ways with hunting, rock-art and the like. The time in the bush would also be used to pass on more stories about their history and culture. Critics occasionally argue, 'What is the point of the traditional bush stuff? It's not going to help them survive in the modern world'. The irony of course is that it does help them to survive—but let's not put 'revenue from art' in the way of a good 'economic rationalist' argument.

> I had never before left a remote town without a full tank, but the next morning I had to.

I said goodbye to Lalla and hitched a lift back to my digs behind the Warburton Roadhouse. While taking in the view from the front veranda of the petrol station later that afternoon, the need for programs such as the arts project sunk in even further. The roadhouse petrol pumps interrupted the view and distracted my gaze. The bowsers were unusual, in that each was housed in a security cage and had a sign attached requesting travellers not to take photos of them. Locking the pumps up was an attempt to deal with the petrol-sniffing problems. The 'arts project' was also partly about dealing with the sniffing problem, as it created employment and gave community members a sense of purpose. I wondered if perhaps it was already achieving that, as I had been told that sniffing had declined over recent months. The decline may have also been because of the ban on unleaded fuel; apparently diesel and avgas, which were both still pumped from the cages, were no good for sniffing.

Gazing at the pumps also brought home my own dilemma—the fuel they delivered was no bloody good for flaming great big German motorcycles. I had never before left a remote town without a full tank, but the next morning I had to.

Navigation by braille

Pulling out of Warburton was the usual cocktail of emotions: a dash of sorrow about having to move on so soon, two shots of excitement over what lay ahead, and lashings of concern about whether I had enough benzene to get there. The stretch ahead was the last leg of the Great Central unleaded famine. Only 240 k's or so down the track was the free running fuel mecca of Tjukayirla. It wasn't a long ride, but I knew it would be a bugger of a push if I had got my maths wrong on the fuel calculations.

A short way out of Warburton my spirits were on the up—a grader was working the road, an omen that hopefully heralded 'smooth times' ahead. For some reason I decided that a photo of the grader would be useful for the website, so I stopped and dragged out the camera. In the outback, stopping

If I had seen a few more of these things bashing around I suspect more of my fillings would have remained where my dentist put them.

in the proximity of an oncoming vehicle is code for 'fancy a chat?', which can be a bugger if you're pulling over just because you're busting for a wizza. The grader driver abided by the stopping code. He pulled up, climbed down and wandered over to me.

Grader Man was in his 60s and decked out in no-nonsense green work wear, a dark shade that merged readily with dirt, perfect for ensuring that laundering was only required every second or third lunar cycle. Grader Man been working the roads of central Western Australia for a long time—his joyless tones led me to suspect too long. His job was a demanding gig that involved months of isolation, week after lonely week of dragging his blade over remote tracks in secluded corners of the country. At the end of each day's grading all he had to look forward to was more of his own thoughts for company. Such solitude could not only do a fella's head in but in some cases his body also.

'Mate, if I'm on a remote track and changing one of those grader tyres and the bloody wheel falls on top of me, I'm stuffed. They won't find out something has happened to me for days', he complained.

It was a valid concern, a concern he'd probably shrugged off when he was 30 years younger and had the strength to extract himself. He went on to complain bitterly about his employer being a tight bugger and not coughing up the cash for a portable satellite phone, a device that would make him feel a little safer. His work wear may have been impenetrable to all kinds of dirt and grim, but his desperation and frustration wafted through it with ease. Trying to be helpful, I thought I would show him some of my safety gadgets. Maybe once he realised how cheap and small they were he would renew hassling his boss to cough up the notes.

'Get an EPIRB mate, they're about the size of a mobile phone. I carry one in my jacket all the time', I told him. I dragged the gadget out and gave him the run down.

'It stands for Emergency Personal Indicating Radio Beacon, or something like that. The point is, if it all goes pear-shaped—like in a life and death kind of way—flick the switch and people from all around the country will come looking for you. Not only does it send out an SOS but the signal also lets them get a fix on your exact location. Plus it has a cool reflector on the back so you can signal passing planes.' His eyes were glazing and I was getting desperate.

'All of that, reflector included, for the grand total of a few hundred bucks', I spruiked.

The talk of money seemed to clear his eyes and push a bushy eyebrow into an arch. I felt encouraged to continue blathering.

'I've been riding around with an EPIRB in my jacket for the last couple of years. If I'm in the middle of nowhere, by myself, and have a massive stack, I have a bit of a safety net. The proviso being that I am still conscious and have a functioning limb to access the thing with.'

Dragging the conversation back to 'accidents in remote locations' had quickly negated the interest sparked by the low price (there are good reasons why I don't work in sales). I was starting to realise that no gadget was going to drag him out of the doldrums he seemed to be in. Years of isolation doing a thankless task had taken its toll. I suspected he'd had enough about 10 years ago and wished he'd been long retired.

As if to put an end to my techno babble he abruptly suggested we both shut down for a bit.

'I'll crank up the billy, we'll share a pot of cha and have a chat', he said. I was also keen to have a chat, but wanted to crank up the recorder while doing so.

At moments like this I hated the job I was doing, being caught in the conflict between the pseudo journalist and the traveller. The traveller wanted to sit around a fire and have a chat with this bloke, but the journo dictated I could only risk spending a wad of time hanging around if I got something out of him. I loathed this cold approach, but I had to have a good reason for risking a delay that may well result in me bashing around in the bush in the dark trying to make my next destination. The unknown conditions of the road ahead were always something that had to be factored in, and his assessment that 'the road is buggered

Just to prove it was not all one big smoothly graded fun park.

after where I've finished grading' suggested I was going to need every spare minute.

Time, or lack of it, was where the compromise always came in. My theory on story gathering was to spend a bit of time with a person and get to know them, and also they you. Then, if they had something interesting to say, ask them if they mind chatting with the tape running. The theory was about building a bit of trust, which therefore required a bit of time. However, as it was already heading towards midday I didn't have enough up my sleeve. I had to go straight for the kill, tell him what I was up to and ask him if he would be interested in being taped as I reckoned he had an interesting yarn to tell.

'Nah mate, I'm not really into that', he replied.

'No worries, I understand', I lied.

For some reason I always took these knock-backs personally. I felt as though I had caused offence. I didn't want people to view me as a parasitic muckraking journo. I wanted to redeem myself in their eyes, but the problem was that all conversation past the point of rejection was hopeless. If I tried to endear myself, tried to convince them that I was not a grubby journo, it just looked like I was trying even harder to get them to consent to being interviewed. The best thing to do is cut the losses and run. This seemed especially wise when I realised how much of a rude bastard I had been—I had not even bothered to find out his name. Little wonder he said no.

As I rode off, ploughing through the earth he had recently turned, I snarled at the compromises I was making. The grader bloke really just wanted someone to have a chat with. If I had taken a few moments, shared a cuppa, it might have brightened up his day. Now, however, he was probably a little darker, convinced I was just a nosey reporter interested in a story and didn't give a stuff about him. I debated with myself, 'What are you, a flaming social worker?' Maybe partly I was. I listen to people, to their stories, with the minor difference being that I retell them to countless other people. Was I being symbiotic or just plain parasitic? Maybe it was just a question of perspective. I was quickly getting wrapped up in one of my internal debates. I knew the signs and decided to shift my focus from my 'date' back onto the track.

One hundred metres on from the last photo I thought the bike might enjoy a 'nice lie down' in the lovely soft sand.

The change in focus was a good plan because the road turned ugly. Central Australia was back into 'bike wrestling' mode. The range country around Warburton was now a long way behind, and the easy going grey gravel roads had turned back into a familiar mix of red earth and sand. The map had warned that this was a sandy section of the Great Victoria Desert—or was it the Gibson? It was a hard to tell—the thin black line indicating the road on my map was drawn through the middle of the two deserts. Despite the mild confusion over exactly what bit of sand I was in, the rest of the details were depressingly accurate, particularly the stuff about the dunes. Usually I try not to consult maps too much—it spoils the surprise. Maps also wreck one of my favourite ways of meeting people … ''Scuse me mate, you got any idea where the hell I am?' However, while I was roaming through a desert (deserts?) I thought that to occasionally open a map might the safe thing to do, even if it did cause confusion over things like which desert I am in. Whatever desert it was, the fact was still the same—the road had turned to crap and I was again working as a two-wheel sand plough, a gig I thought I had left behind days earlier. To complete the sense of deja vu, I decided another photo of yet another sandy road would be a good plan. Without missing a beat I repeated the stuff-up I had performed when I first got on the Great Central: stop in the sand, get a photo, try to get going again, fall off. Most people would try and cover this sort of stupidity up. I, however, thought it would be a good idea to share it with the on-line world and grabbed another set of snaps of the upturned bike for the website.

You can call the Great Central all sorts of things, but certainly not monotonous. What you are looking at is a bloody great big water hole in the middle of a desert, contrasted next to it is a bit of desert scrub that had been recently vaporised in a bushfire.

After much grunting, swearing, sweating and time I righted the bike and was wobbling off through the sand. My efforts to deal with the sand were being thwarted by the rear tyre I had fitted in Alice Springs. I hadn't been able to get the brand I wanted—I was forced to go with the 'B' option and now I was suffering for it. It felt like it belonged on a shopping trolley. The tyre flexed and twisted, urging the bike to veer towards obstacles like a trolley would go towards an overstocked shelf. To add to my frustration, the early afternoon light demanded a new riding style—navigation by Braille. The desert sun hit the track at the perfect angle, turning it into a featureless red sheet. A fierce glare kicked up off the sand and obscured all the contours of the road surface. Wash-outs, bumps and holes were discovered after the event as I bounced blindly along the track. The bike, however, did not seem to be bothered by the minor deflections in course; if anything, it wanted to go faster. Eventually, grudgingly, it settled for speed of about 110, but for such terrain it still seemed a little quick on a big bike carrying a full load. I tried to take some comfort in the thought that 'if it went wrong, it would go spectacularly wrong'. I doubted, however, that a disaster was going to happen—the bike only seemed to punish me with a fall when I went too slow—so I resigned myself to a whiplash tour of the desert and clung to the handlebars.

The road continued to carve through the dune country, hemmed in on either side by low scrub and tufts of grass. The vegetation occasionally vanished, consumed by recent bushfires that left the sand exposed, a red nakedness blemished only by occasional piles of white and black ash. The torching had been so complete it looked like someone from the Starship Enterprise had popped down for a day of vaporising scrub.

The afternoon became a juggling act: marvelling at the landscape (in frantic glimpses), cursing the tyre, ricocheting off unseen obstacles and recalculating my fuel situation. Fuel became the most concerning part—one moment I was convinced I would end up pushing the bike, the next I was sure I had heaps left. This seesaw of hope and despair, pivoting around my dodgy calculations, made the k's turn over slowly. The odometer began to roll over a little faster when the sand eventually gave way to a hard packed ironstone road. The more the number tumblers quickened their pace the more confident I grew about romping it in on the fuel front.

As I continued my journey across the Centre, one theme remained constant—the abundance of healthy vegetation in what I had expected to be barren deserts. Whenever I commented to anyone about this, I was quickly reminded that the country looked so good because of 3 years of good rainfall. That point was rammed home one last time when I was about 30 k's short of my home for the night. A massive waterhole lapped up to the edge of the track—if I had been a few days earlier I would have spent some time waiting for the water to go down and the track to clear. I had not seen a decent body of water on the whole trip; a couple of shrivelled up dams but certainly nothing that would take a good hour to walk around.

Waterholes in the middle of the desert are a rare thing so I imagined it would be teeming with birds and animals, and that classic Kodak moments beckoned. I climbed off the bike, slipped into my best Dave Attenborough, and crept off around the edge. After half an hour of enduring my own hushed tones I had seen stuff all. In fact, the only worthwhile photo would have been of me floundering around trying to liberate myself from knee-deep mud. Extracted and squelching, I waddled back to the bike, musing that only a select few people could visit an oasis and get a photo of nothing. To reinforce my lack of 'wildlife credibility' a small swarm of wasps took a liking to the bike in my absence and were hostile about the notion of sharing it. Being none too

L to R: The RFDS landing strip at Tjukayira's, the only bit of the Great Central Road that is bitumen.

After 1200 k's of sand, ruts, corrugations and dunes I had my doubts about the 'easier' claim.

partial to wasp stings I decided to wait quietly, hoping they would bugger off. No joy. The next option was to up the ante: I got a big bushy branch and charged the bike. No joy again. In fact, perhaps not surprisingly, they were pretty annoyed about my advances and I was forced to leg it and wait for things to calm down. The final solution was repossession by stealth. I moved slowly to the bike, gently climbed on, delicately put my hands on the grips, flicked the start button and rode like buggery.

After 30 k's of riding one-handed, patting and searching my suit for wasps that avoided the high-speed getaway, I was staring down at Tjukayirla roadhouse. The roadhouse had an impressive welcome mat out the front—a couple of k's of tarred runway, complete with painted white bits at either end. The runway also doubled as part of the Great Central Highway, which was convenient for the roadhouse as it helped keep the dust down. I had read somewhere that Tjukayirla's claim to fame was 'most isolated roadhouse in Australia'. I had also heard that Rabbit Flat roadhouse in the Tanami desert claimed the same crown. Quite frankly I didn't care. It was clean, tidy, dust free and offered food, accommodation and, most importantly, unleaded fuel.

While I was paying quiet homage to the petrol pump, a couple of 4WDs pulled in. Both were white, covered in dust and pulling trailers. Both Cruisers had roof-racks with tin signs bolted to them proudly announcing the 'Brian Young Show'.

I soon found out that the 'Brian Young Show' was a band. Out of the 7-piece band, Brian was distinguished by not only being the leader but also by being the shortest and the oldest, by a long way. Brian looked like he was comfortably in his 50s, while the rest of the band members were in their early 20s. On top of Brian's head was a broad-brimmed Akubra, the front of the brim bent sharply down, obscuring all features north of his nose tip with an impenetrable shadow. Southwards was a broad and ready smile, supported by a chin-cum-neck that eventually vanished into a blue shirt. Central Australia was Brian's office and I had bumped into him between jobs.

'I tour Australia with a 7-act show. I've been doing it since 1977. We tour all the remote areas, right out in the desert, as well as all the cities, big towns, small towns', he told me proudly.

Each year Brian spent $3\,^{1}/_{2}$ months doing 35 000 k's of travelling. He used to tour the nation by plane, but a crash and escalating costs forced him to resort to long k's on the road. A couple of decades travelling the bush had resulted in him gaining unique access to some of the communities and country.

'When I first started, a lot of them were little tiny kids. Now they are grown men and women. It's strange, because they believe I should know their language seeing I've been coming out for so long, but the problem is I'm never in the same place long enough.'

Even though his visits to the communities were brief, Brian was the only performer who regularly made the effort to visit them, which resulted in him being held in unique esteem.

'I've been to places when there was men's business, sorry business, and normally people are not allowed, but they say "you make our kids happy you come through." In some cases I'm not allowed, but I certainly get a lot of privileges that a lot of other people wouldn't', Brian said, in an understated way.

I put it to him that he had worked damn hard to get those privileges. He seized on what I said, keen to correct any possible perception that he was being dismissive.

'You've got to do the right thing. You come out into this country, the desert, and you do the wrong thing by their laws, you'll never get back in again,' he said.

For audiences to keep on embracing him, year after year, must have meant that he was doing something that they liked.

In the middle, wearing a white shirt, is a bloke by the name of Brian Young. Brian is a bit of an Outback legend because each year he and his band climb into a couple of troop carriers and set out on a 35 000 kilometre bush bashing expedition to play the remote communities throughout Australia.

'A lot of people believe that because I take the show out to the communities it's got to be a hick show, and it's not a hick show at all. I mean, I've got all these young acts that sing country, country rock and old rock and roll. Out here in the desert you cannot do a straight country show because they would all just pick up their blankets and go home. They want old-time rock and roll, they want the old four-four beat because that is their dance beat. That's originally their corroboree beat—when you start playing that type of music they all bounce up and dance. It's great to see.'

I was expecting a chat about play lists, but this was far more interesting.

'Really, you're just tapping into a natural rhythm?' I asked.

'Oh yeah, you got it. I wrote a song called 'Big Fella Wudumbah', which is based on an old Aboriginal story where they used to say, "Don't worry about the white fellas bringing in their cattle and horses, one day a big flood will come along and wash them all away." An old elder up on Cape York said, "You know why they like that song?" and I said "no". He said, "That's their dance music, that's the way they dance."'

With that Brian broke into a few lines of the song. Being 'musically challenged', I was struggling with the 'four-four' part. When he finished, Brian suggested that if I wanted to get a different perspective on his bush tours I should speak to one of his band members, Leigh Forster.

After chatting to Leigh for a bit it didn't take me long to work out that being in the band was viewed as a privilege. Sure it may not be headlining the Big Day Out, but for those chasing a country music career, doing a tour in the Brian Young Show was a big leg-up.

'If you complete the Brian Young show it's a little bit of an apprenticeship. It's a hard show—you're with 7 people in 2 troop carriers for 4 months. You're at each other's tucker boxes, living out of swags—you've got to be pretty friendly. You know what it's like when you're on the road', Leigh said, nodding towards me.

'Plus you've got to play each night, the same song, but you've got to keep up the standards and be fresh because the crowd doesn't know what you're going through each day.'

To reinforce how important a rite of passage the tour was to young musos, Leigh went on to detail what happened if you stuffed it up. 'People have been kicked out. If that happens you're shamed and find it hard to get anywhere after that happens.'

I was a little surprised to hear this.

'Brian's show has that big an influence?' I asked.

'Sure. Once we hit the east coast, a lot of industry people are looking at who's on the show. I mean they've had Troy Cassar-Daley and Beccy Cole—some of the big names have been on the show,' Leigh told me proudly.

Besides the career move, the main thing that seemed to excite Leigh was that he was travelling through the outback playing country music.

'I just think touring and playing country music is what it's all about. Just getting out in the bush and seeing where it's at', he concluded.

With that I let the crew get on their way. They, like me, had commitments and a budget to adhere to, so they had to keep moving. Although my budget and timetable occasionally made me want to scream, I tried to look at the flip side: I was in a unique position of mixing travel with work.

The concept of work, though, was beginning to weigh on my mind, largely because I hadn't done any for a while. I had gathered a lot of material, but because I had been staying in dongers for days (transportable huts featuring only a bed and grime) I hadn't had the space to set my equipment up and begin editing and producing stories. The notion of getting further and further behind weighed on me as I headed to my bunk for the night. The room I had at Tjukayirla was called the Hilton and it lived up to its flash name by featuring a bed and also a desk. For the first night in ages I would be able to resort to my normal evening pattern of working until I fell asleep at the laptop. At 10 pm my plan was thwarted by management—apparently they saved money by turning off the generator at that time each night. I climbed into bed wondering if Mission Central would understand that I had not filed anything for a while because simple city commodities, like desks and electricity, weren't always easy to come by in the middle of the outback.

Reptiles and discount fuel can do strange things for a lonely man in a desert.

WALLY AND ELSIE

'I'm on the wrong side of 75 and haven't got many motor-biking years left—well, I don't think I will have anyway. So while I am able to go, I'll go for it', said Wally.

Wally and Elsie for their 40th wedding anniversary decided to bypass the romantic candle-lit dinner concept. Instead they chose to celebrate by climbing on to a bike that was manufactured the same year as their wedding and riding two-thirds the way across Australia.

For a couple of septuagenarians, Wally and his wife Elsie were going for it in a big way—they were riding a motorcycle from Bunbury in Western Australia to Streaky Bay in South Australia. In anyone's book it is a reasonable haul. In Wally's book of imperial measurements it was '1200 miles—one way!' I met Wally and Elsie in the middle of the Nullarbor Plain. They were halfway through their journey and I was on my first trip around Australia. Wally and Elsie stood out not only because of their age but also because of the bike they were riding.

'The bike is an old ex-police W1 Kawasaki with a copy of a Rudge sidecar attached', Wally said proudly. The important detail to distil out of this was that it meant the bike was about 40 years old. Admittedly it wasn't quite the same vintage as Wally, but it still lacked many of the conveniences of a modern bike—an electric start was one of them. Each morning Wally would have to kick-start the beast into life, which was a pretty impressive feat for a bloke in his 70s. I was even more impressed with Wally's efforts after he told me the motorcycle trip was their way of celebrating their 40th wedding anniversary. Most people commemorating such a milestone would probably do something a little more sedate, luxurious and comfortable, like an ocean cruise. Elsie, though, seemed very happy about the bike option.

'It's very, very comfortable and very interesting as well. You get a hands-on view of the country instead of looking through the window of a car. It's a different view altogether. You sort of get an insulated view in a car—you get the real thing on the bike.' Elsie waxing lyrical about the bike had an added dimension when I found out she had only recently become involved with bikes.

Nearly 80, in love and getting around in something far more exciting than a motorised wheel chair. How many septuagenarians can put that on their CV!

'My husband has always been interested in motorbikes since he was 17. Me, I've only been involved for a couple of years', she said, as though it was quite common for people to begin their involvement with motorcycles in their 60s. Then again, maybe it was. After all, my own mother had waited until she was nearly of that vintage before chucking a leg over a bike.

'My husband has always wanted to go across the Nullarbor, so I said I would certainly come if I could go by sidecar'.

I thought it remarkable that after 40 years of marriage they were still having new experiences together. I asked them what their friends said about them riding an old motorcycle across the Nullarbor. Wally took up the baton:

'Well, I tell you what they say—they reckon I want certifying and putting away,' he rhymed.

'What do you say?' I asked

'I just laugh. I don't take much notice of it.'

Despite his age, Wally looked pretty damn healthy. I had noticed the tone of his arms when he was manoeuvring his bike—I would have thought twice about an arm wrestle. My guess was he needed to be especially fit, as the bike and sidecar looked like a heavy, cumbersome rig to grapple with each day.

'Actually it is. You've got to watch it all the time, but you get used to it', he said seriously. He went on to point out that because of the size, and the small motor, he stuck to 45 miles per hour. They certainly weren't getting a whiplash tour of the Nullarbor, especially considering Wally's travel plan: 'We aim to do about 120 miles a day'. This mileage didn't sound like much, but when you considered their age and the weight of the bike, then it was probably a fair day's ride. Wally gave me a tour of the bike and I discovered he was carrying a heavier load than I'd thought initially.

'I've got a spare battery, clutch plates, chain, anything I might need', Wally told me.

'And you'll be able to do all the work yourself?' I asked.

'Yep, you've got to, especially out here.'

Wally and Elsie were an inspiration. I knew people one-third their age who were less capable and far less active.

'Before I retired, there were several other blokes that retired, and they just sort of shut down, and some of them only lasted 18 months', recalled Wally. 'Yet while they were at work they were like dynamos, full of beans. When they retired they went downhill and lost all interest in life. Work was their hobby and their life—they couldn't see anything else. There was no reason they should have died. I thought to myself "I'll never get that way I hope", and I've made sure I never.'

Wally made sure he never by having a lot to go on with after retirement. He had a simple philosophy on it.

'That's what keeps you going. Just sitting down in front of the TV all day, you wouldn't last too long. You want to keep yourself active', Wally concluded.

Wally and Elsie had to get on with it and cover their 120 miles for the day. Elsie packed herself into the sidecar, Wally climbed onto the bike and with a single deft kick it chugged into life. He gave a toot on the horn and I waved the two lovebirds off. As they faded into the distance I couldn't help but feel I had been very lucky to meet them—I now had something to aim for when I got to my 70s.

15
Tar school

In central western WA, outside a town called Laverton, the Great Central Road wrapped up with a whimper as its sandy surface merged uneventfully with a bitumen road. At this point of conclusion there was no sign bestowing congratulations for successfully navigating 1200 k's of dirt, sand and corrugations, just a featureless narrow blacktop that meandered into the town. Despite the lack of a welcome committee I was chuffed I'd made it and decided a photograph of an intact 'bike and Drew' was appropriate. I parked the bike where the sand and tar blurred together, erected the tripod and set about getting some happy snaps. Mid photo shoot a bloke in a council truck drove past. He rolled his eyes in a manner that suggested every day he saw some dickhead parked at the end of the Great Central taking photos like they'd just conquered Everest. (His expression may

L to R: Things to get excited about in central Western Australia… The end of the Great Central and long distance telephones

equally have been one of disbelief that some clown would park in the middle of the road to take a self-photo.)

Despite the luxury of tar, Laverton had little to keep me captive. Even if it did, I didn't give myself a chance to find out. I was keen to push on to Kalgoorlie. I had a commitment at a radio station there and was also desperate to bunker down and start filing the stories I had gathered.

Kalgoorlie amounted to a blur of days in front of the laptop, churning out the goods, by the end of which I was hanging out to up stumps and get back into the bush. I rode out of town with a vague plan of heading north-west(ish) to a dot on the map called Sandstone. The map suggested if I went straight north I could go most of the way on bitumen, but where was the fun in that—especially if I could get there on a dirt road.

The track for Sandstone started at a town called Menzies, a place that seemed to be renowned for its dead. Menzies, from what I could work out, got off the starting blocks when someone mentioned 'gold'. The gold rush came and went, as did the bulk of the population. The town's present day claims to fame appeared to be STD phone calls and a cemetery. The STD phone facilities were proudly promoted with their own roadside sign at the petrol station. Surprisingly, though, there were no signs advertising the cemetery, which, thanks to the gold rush, was probably one of the more unusual ones to be found in the Australia.

Menzies in central Western Australia was once home to a multi-tasking blacksmith. However if your family couldn't afford his services then you could look forward to having your life summed up with a mound of dirt and numbered peg.

Judging by the graves, surviving a gold rush was a bit of lottery. I began to suspect that being an undertaker was probably one of the more stable forms of employment in early mining towns. While Menzies may have had an undertaker, it seemed they lacked a headstone maker, a problem that was overcome by a bit of bush improvisation. The town's blacksmith was put to work and new cemetery residents received a tin headstone. (Coupling the word 'tin' with 'headstone' admittedly is a little contradictory. Unfortunately 'tombstone' and 'gravestone' face the same problem. 'Headtin' seems pretty dopey, and 'tin grave marker' a little inadequate, so 'tin headstone' or just plain old 'headstone' it is.)

The tin headstones were the same height and width as a normal headstone, the differences being that they were 3 times as deep and looked like tiny houses, each one complete with a pitched roof, gutters and a glass front. Inside each headstone were the remnants of artificial flowers and an epitaph. Above these, nestled in the gable, was the date of death. The headstones were finished with tin lacework on the front edges. The really flash graves were distinguished by having edging around the plot, which were essentially little fences made from galvanised pipe. The truly poor, however, were not afforded the luxury of headstones or pipe. One section of the graveyard was heavily populated with bare mounds, each marked by a cast-iron peg with an embossed number. The scrub near the 'semi-marked' section of graves was gradually encroaching. It wouldn't be long before the bush would reclaim them.

I visited the cemetery to see the tin headstones and left remembering why I avoid graveyards. Inscriptions of young deaths from 'consumption' and 'mine collapses' coupled with rows of unmarked forgotten graves were not a great celebration of lives lived. People carp on about the 'good old days', but I doubted whether many of the people planted there thought it was all that good.

I climbed back on the bike, gave the throttle a flick and rode off north-west, contemplating that I wouldn't be heading west for much longer—I was running out of continent. In a way I was happy about this. I had been pushing west for a few thousand k's and I'd had enough of the afternoon sun drilling into my eyes. Sun glare, however, was not part of the equation on this day. The sky was black and pregnant with moisture, the flotsam of a tropical storm that had drifted south. It was nice to be out of the heat, but I wasn't sure if swapping it for rain was the best deal I could have hoped for—a good downpour would turn the track into a quagmire. For half an hour fat drops splattered intermittently against my jacket, constantly threatening that the sky was going to soon get serious about delivering a mother lode. I was confident that if I stopped and went to the effort of putting my wet weather gear on, the rain would be sure to bugger off, so I wound down through the gears and parked on the side of the track. Getting waterproofed required stripping out of the suit and zipping in some Gortex liners. I had only seen one vehicle in the last couple of hours so I felt reasonably confident that I could strip down to my duds without being interrupted. Naturally, when I got down to my boxer shorts a truck appeared. If it had driven straight past it would have meant enduring a brief moment of embarrassment, but this bloke decided to rub salt into the wound and pulled up for a chat.

'Getting a bit wet, isn't it mate?', asking and stating the obvious in one hit.

I thought it best to seize this opening and explain to the truckie why I was standing in the middle of an outback road *sans* duds.

'Yeah, it has come in a bit. I figure if I make the effort to stop and put in my waterproof linings it will be bound to bugger off', I proffered.

No wonder these things get mashed all over the roads of Australia, how the hell is a fella supposed to spot this while doing the metric tonne? The worrying thing is that on a motorbike the 'mashing' process is a two way deal.

'Dunno mate, not sure what she'll do. It could get real wet', he forecast, oblivious to my attempt at humour.

'I guess we'll both find out as we get further up the road', I replied.

With that, the conversation was pretty much over. He wished me luck and drove off. I wasn't sure if the point of the conversation was to make me squirm or if he was just a friendly bloke who encountered trouserless people everyday so it was no big deal. Whatever the case, I was happy that the truck was ahead of me; I would hopefully be able to follow in his wheel tracks if the going got slippery.

The rain continued on in a gutless, dribbly fashion, the drops meandering across my visor, blurring my vision. Poor visibility on a bike is never helpful, but coupled with the pale light limping down from the black clouds it made for tenuous travelling. The half-light also confused the crap out of bush critters—they thought it was dusk and started climbing out of bed and bounding over the road. Poor visibility, a greasy road and early rising marsupials—the day was going perfectly to plan. I was looking forward to getting to Sandstone.

Eventually, after one last close call with a wallaby, I was in the metropolis of Sandstone. The town featured a post office-cum-general store, a petrol station, a pub and so on—all the bits and bobs that make up a small regional town. The thing that set it apart from most other towns was that Sandstone lacked tar: the place was bereft of bitumen. The red earth began where the mortar of each building finished making the town look like it had been dumped in the middle of the scrub. The few people wandering the street all stared at the sky whenever they stopped to talk to each other—weather conversations seemed to be big. I guessed this was understandable. Sandstone not only had dirt streets but was also connected to the rest of the world via dirt roads and a dirt airstrip, so heavy rain would always be a worry and a hot conversation topic. It didn't take long until I found out how much of a worry it really was.

'There has been a couple of times when the Royal Flying doctor hasn't been able to come in and land because we've had rain. We also haven't been able to get out because all the roads are closed', Beth said.

Beth spoke with a degree of authority, firstly because she had lived in Sandstone all of her life, and secondly because she was the Shire president, the school registrar, the golf club president and the local storeowner. (Outback folk had 'multi-tasking' down pat a long time before some management spiv coined the phrase.)

Despite the problems caused by the roads, Beth wasn't too fussed about it. This was probably because the years of weather anxiety were about to be put to an end.

'We are about to get a tar road that will run from Mount Magnet, through here and on to Linster. That's going to make a huge difference to the town. There will be less dust getting in people's houses and it will be much easier on traffic, especially trucks carrying goods to the town. So it's a pretty exciting time.'

Seeing someone get excited over 'tar' was not something I had experienced before, but after having the problems of 'dirt' detailed I was beginning to understand why. Safety was just one of the benefits the humble blacktop would bring; the locals hoped there would also be a reduction in the cost of goods as trucking companies would no longer have to build in a wear-and-tear premium for traversing the rough dirt roads. The tar also brought hope—hope that Sandstone would grow, as opposed to shrink, like so many other rural centres. The completed loop between Linster and Mount Magnet would mean that trucks travelling between the north-west and the south-east of Australia would soon be taking a short cut through Sandstone. This would bring cash from passing vehicles and hopefully it would also bring income from tourists—Mr and Mrs Grey Nomad with their camper trailer hooked to the back are not renowned for venturing beyond the bitumen. There would be no excuses now.

Apart from injecting some new sources of revenue into town, the arrival of bitumen would also save them some cash in the long run. Although bitumen roads are not cheap, they are low maintenance once in. Dirt roads require continuous upkeep.

L to R: Sandstone was yet to experience the joys of tar, which for the locals meant dust was a never-ending pain in the quoit. The problem was partially addressed by this beastie, it roamed the streets and surrounding roads each day, dumping 350 000 litres of water to dampen the problem.

The hustle and bustle of downtown Sandstone.

'If a truck comes through on a rain-closed road it just means it tears up the road and the Shire has to go out and do a lot of work to fix it up and that's a lot of dollars', said Beth.

With all the benefits of tar listed off, I left Beth to return to her job of the moment; it was school time and she had registrar business to attend to. I climbed on the bike and rode to my digs for the night, a huddle of portable cabins on the outskirts of town.

The entrance to the cabins was blocked by a semitrailer, a rig that I found out was soon to become redundant because of the town's blacktop. The truck was a water tanker that sprayed the roads to help keep the dust down. It also helped soften up the soil for a grader to occasionally knock the corrugations out. The water tanker looked like it was originally white but over the years it had been turned into a rusty red, a change that was courtesy of a combination of red dirt and mineralised water. Everything in the west seems to be tied in with minerals and the bore water was no different; in places it is heavy in iron and leaves a rusty stain on everything it touches.

Tinkering around in the truck cabin was a bloke caked in as much grime as the truck.

'I'm dumping about 350 000 litres of water on the road each day', Craig stated, sounding bored with the number, like a tour guide becomes with a fact they've said 100 times too many.

'That's a lot of baths mate! How many trailer loads is that?' I asked, trying my best to get him to chat in a bit more detail.

'Only about 10', Craig mumbled. My enthusiasm was failing to pay dividends. I suspected being economical with words was Craig's way, and to be fair, I was also terrorising the poor bugger with a microphone, a device that has an astonishing ability to make people either rabbit on like idiots or clamp right up.

I was 'Tonka deprived' as a child, which has manifested itself in an endless fascination of big things as an adult. For Craig, this meant I wanted to know everything about the truck and also see it in action. Mounted at the rear of the tanker was either a plumber's triumph or nightmare: fire hydrant-size pipes crisscrossed each other as they snaked out from a pump and terminated in an array of large cylindrical spray heads. Craig said the plumbing duct-work delivered all 35 000 litres on the ground in less than 20 minutes. I was itching to see it in action, but I suspected Craig was growing tired of my endless questions so I was reluctant to ask him to fire it up. Fortunately, though, he was doing some maintenance and had to start it anyway, warning me to stand well back before he fired up the pump. I took his advice and found a spot about 5 metres back.

'Nah mate. I said "stand back". Keep going', he urged.

Twenty metres later Craig deemed me to be out of firing range and he pressed the start button. The engine roared and a horizontal torrent leapt from the spray heads and rushed towards me. Fortunately the flood fell short of me, but not short enough for my legs to escape a splattering of mud as the water ricocheted off the red dirt. As I stepped back to gain a bit more clearance I appreciated Craig's earlier directions—if he had wanted a laugh he could have left me where I was and watched me get blasted 10 metres through the air and land on my butt.

The coming of bitumen was going to make Craig one of the few losers, as the tar would put him out of a job. Craig, though, was philosophical about it.

'There's plenty more places that don't have tar. I'll just have to move further inland for work.' Despite the stoical comment, I suspected he wasn't all that happy about it. His wife and children lived in Mount Magnet, a couple of hours to the west, which meant he had to live in Sandstone during the working week. Soon there would be even more distance between him and home.

Craig returned to working on his truck for a bit before concluding that the switch he was trying to fix was buggered. The switch operated the spray pump and without it he would have to spend the day getting in and out of the cab to manually turn the pump on and off. He shrugged his shoulders, climbed back up into the cabin and drove off, winding his way into the distance through a myriad gear changes.

While Craig lost out from the arrival of the wondrous black strip, there was a small group (28 to be precise) that it didn't make a difference to at all. The students of Sandstone primary knew that, come the end of Grade 7, their schooling options would remain the same as always—leave town and go to boarding school. Sandstone did not have enough students to warrant a secondary school and the coming of tar was not going to suddenly boost numbers any further. (As it was, they thought they had done remarkably well in regards to growth because student numbers had nearly doubled in recent years. To cope with the influx a second teacher had been employed and the school was now split into two: lower primary and upper primary.)

Filling me in about the pitfalls of outback schooling were Karen and Steve. They had a vested interest in the system as Adam, their 12-year-old son, was in his last year at the school and would be soon heading down to Perth to go to boarding school. Karen viewed the topic as a bit of a case of swings and roundabouts.

'You do have your restrictions in a small town. Because we don't have enough children to make up any team sports—there's no footy, there's no cricket or anything like that. The kids do individual sports like tennis', she told me.

Sandstone primary was booming when I passed through, student numbers had reached 28 and the school had been forced to boost teacher numbers from 1 to 2. Adam, pictured, was in grade 7, the end of the line for his education at Sandstone. Year 8 was going to be a whole new world of boarding school a long way away in Perth.

While moving to Perth provided Adam with more opportunities, like playing team sports, it turned out he didn't have to go there to continue his education.

'We do have another option here at the school, which is distance education through the secondary level. Some children are adaptable to that and some children aren't, but 99% of the time you just have to accept that your children have to go away to boarding school to get the education they need for secondary level', Karen said.

'Does it worry you that it spells the end of the family unit as you know it?' I asked, a little more bluntly than planned.

'Oh, most definitely. We've got a trip booked in a month's time, and this will be the last holiday as a family. Adam will come home 4 times a year, 2 weeks at a time, as opposed to him being here full time', explained Karen.

Not satisfied with her only baring her soul that far, for some reason I decided to push on further.

'How do you feel about that as a parent?'

'Well, when you move to these areas you just accept that's the way it's going to be and you just prepare yourself the best you can, you know, mentally and emotionally, the best that you can.' Karen paused, then added, 'It'll be hard, yeah'.

If Karen thought it was going to be hard then I probably should have spared a thought for Adam, after all he was the one who was about to be dumped into an alien environment. Adam was leaving a small community where unknown faces were rare, a place where there were few 'stranger danger' fears. In Sandstone, Adam had a huge amount of freedom. He could disappear for hours on his motorbike or he could jump in a ute by himself and do an 80 k round trip to check all the bores. At the right time of the year Adam would head into the bush to stalk emus and raid their nests for eggs, which he sold for pocket money. Every day was an adventure and it was largely because of the space and the lack of risk from strangers. I suspected that the city was going to be a bit of an adjustment for him.

The move to the city was still a year away, and the family was making the most of the time they had together. Part of that included a little ritual Karen and Steve observed every day—they would stop and watch the sun go down.

'If you take the time to do it, you'll be amazed. No two are ever the same', Stephen assured me when I joined them.

It was something I had never considered, but as I watched a cascade of crimson I realised he was right—they are never the same. Sandstone wasn't going to be the same either after the tar came through, in ways that perhaps the locals hadn't bargained on. For all the positives, there was potentially going to be one great loss—freedom. The freedom that Adam and his mates took for granted was going to change when road trains, traffic and tourists began to haul through. The tar was going to deliver lots of unfamiliar faces. As the sky went through another shade of crimson, I couldn't help but think sadly that growing up in Sandstone might become a little less of an adventure for the kids around town.

Random Snaps from Lap 1 L to R:

Milking the bike for extra fuel near the end of a 400 k no fuel section in the Barkly Tablelands—leaning the bike over gets a bit of extra juice across to the fuel pick up.

Dinosaur country, Hughenden Central Queensland.

The shire with no town

To get to the Pilbara I wanted to head north-west out of Sandstone, up to Meekatharra and through the Gascoyne and then into the Pilbara. That was a nice plan but it was foiled by commitments out west in Geraldton. The first of these involved visiting the local ABC station and chatting on the wireless; the second concerned collecting a tyre. As much as I enjoyed the radio chats, the new rear tyre was the thing I was excited about. The tyre I had been on for the last 3500 k's was, to be generous, a piece of crap. The sidewalls flexed in all the wrong ways and the rear of the bike moved about unexpectedly—fun for half an hour but confidence sapping for half a continent. I had a good mind to call the guy who had sold it to me and tell him to revise his sales pitch. His chant of 'this brand of tyre is way better than what you've currently got on' had rung in my ears for the last 3000 k's.

Accompanying his words was the drone of the flaming thing—it was the noisiest tyre on the planet. It sounded like I was being chased by a swarm of bees. All in all, the tyre was doing my head in and I was hanging out for the remedy.

Despite the anticipated tyre relief, my obligations in Geraldton weighed on me, largely because I was not happy about breaking up the trip. I would be leaving the outback and heading into the populated coastline. With this in mind I was keen to get in and get out as fast as possible.

Sandstone to Geraldton was approximately a 600 k hop. That was a reasonable day's ride so I climbed on the bike early and set about getting the transit over and done with. Unfortunately the morning ended up being a case of the 'best laid plans' going astray, as I had overlooked the implications of the new tar road being built and got to spend a couple of hours weaving through 100 k's of road works. My navigation around earthworks, trucks, graders and bulldozers was set to an ongoing backdrop of overcast skies. The heavy clouds left me unsure of whether I should put wet weather gear on, and the wildlife was muddled about whether or not to be in bed: judging by the numbers that ambled out in front of me I suspected sleep deprivation was beginning to take its toll.

> Considering I was supposedly in the communication business, the lack of a phone made for interesting weeks ahead.

Heavy skies accompanied me until I was nearly at Geraldton, then the sun decided to make a point of breaking through as I crossed into that green belt that hems much of Australia's coastline, that safe zone which nearly 80% of us live in and understand. The rolling green hills, lush with healthy grass and packed full of grazing livestock, reminded me of home. The thought was reinforced further when the hills eventually plunged down into the ocean; the 'deep blue' was something I had not seen in 5000 k's. Strangely the sight of the coast put to rest my anxiety about leaving the outback; it seemed appropriate that the western most point of my journey should be defined by me laying eyes on the Indian Ocean.

My stop in Geraldton amounted to a few frantic days of changing tyres, engine oil, talking on the wireless and banging on the laptop, trying to catch up on work. My last task, which I thought would be a nice gesture to the people I was going to meet over the coming weeks, was to wash out my stinky riding suit. Altruism has a lot to answer for because unfortunately I forgot to take my mobile phone out of the pocket before dumping it into the wash trough, so I left Geraldton *sans* mobile. Considering I was supposedly in the communication business, the lack of a phone made for interesting weeks ahead. Sure, it only worked in

Considering I was supposed to be bashing around the Outback, a seaside stop seemed kind of inappropriate, but commitments in Geraldton demanded that I made the 600 k detour…The flipside was that laying eyes on the sea was perhaps fitting as it marked the western most point of my journey.

big regional centres, but it was bloody useful when it did work. The flip side was that being without a mobile seemed kind of appropriate. After all, most of the people I would be meeting in the bush didn't have one either.

Spending a few days sorting things out in a large regional centre like Geraldton was not unlike stopping in big towns on my ride back from England. In countries such as Pakistan, only the big cities like Islamabad had the facilities to repair broken bits, banks to transfer money and phones to order spare parts with. The comparison, though, is largely a romantic one. Australia has infrastructures that developing countries can only fantasise about: ever-improving roads and communications are shrinking the outback. People blighted with a breakage in the bush need only limp or get towed to the next dot on the map and chances are that help can be organised and spares freighted in within a few days. Finally, when replacements do turn up (as opposed to 'if' in developing countries) there aren't language barriers and the unwritten rules of baksheesh to wrangle with.

Returning to the outback involved riding out via the route I had come in. The green hills eventually gave way to a dry, sparse countryside, and 100 k's later I hung a left in a town called Mullewa and began my push northwards to the 'Top End.' The leftie put me back on to a dirt road, which to me symbolised heading into the outback proper again. In case I was in any doubt about this, the locals, like those in Port Augusta, had put a sign up:

'Welcome to the Shire of Murchison
Where the Outback begins'.

Murchison is anything but alone in claiming a title like this; in fact, it's so widely abused that even one of the slogans on SA numberplates reads, 'South Australia—Gateway to the Outback'.

Claims of being 'The start/beginning/gateway to the outback' had reached epidemic proportions and the folk of Murchison were considering changing their badge to:

'Welcome to Murchison,
The shire with no town'.

The shire covered some 40 000 square k's but had only a 150 residents. With so much space and so few people, the Murchisonians decided not to crowd themselves into the one location; consequently the shire had no town.

The closest thing to a town was the Murchison roadhouse, which was also going to be my home for the night. Unfortunately, though, cancer had different ideas about that. The manager of the station was ailing with the disease and the roadhouse was temporarily closed. I felt for him and his family but I also felt selfishly for me, as it meant no food, no fuel and no bed.

> One of the clichés about the Outback that does ring true is that when things get tough everyone mucks in.

One of the clichés about the Outback that does ring true is that when things get tough everyone mucks in. In the case of the roadhouse it involved handing the keys over to the neighbours down the road. In a chunk of land the size of Murchison, neighbours are often half a day's drive away. Fortunately this time the neighbour was only a few hundred metres down the track.

Neal, with wife Elsie in tow, had moved to Murchison to take over the reins of the Shire. The CEO gig was to be Neal's last challenge before they moved down to Perth to retire. The striking thing about Neal and Elsie was that for a couple who weren't born-and-bred locals, they were remarkably proud of their adopted home. Neal was particularly chuffed about the fact that he was probably the only shire boss in Australia that didn't have a town in the middle

The legacy of Monsignor Hawes. Spanish inspired architecture plonked in the middle of the outback by a guy that was born and raised in England… it makes about as much sense as a tree in the middle of nowhere full of stuffed toys.

of his patch; so proud in fact that it was he who was toying with the idea of the new shire slogan.

Because Neal was running both the shire and the roadhouse, I imagined he would be a little short on time. Accordingly, I expected my meeting with him to be a brief 'g'day, here's the key to your donger' type thing. One of my travelling mantras, that I have always failed to live by, is 'don't have expectations'. When I turned up to Neal's front door I was greeted with, 'G'day Drew, forget the dongers, you're staying in our spare bedroom and dinner is at 7 o'clock.'

Being offered hospitality is one of the great things about travelling, but on this occasion I was a little apprehensive, as it would involve me traipsing through their nice home with all my grubby luggage and riding gear (the recent laundering had only been good for about 300 k's). As well as my fear of imposing, there was also the issue of needing to work each night. I was caught in a never-ending game of trying to catch up with myself. I was still filing stories about the Great Central Road of nearly 2000 k's ago—a night off meant the stories would just get another 300 k's behind where I actually was. I could hardly stand up

> As I sipped another cold beer and soaked up the same TV programs that the rest of Australia was watching, it was easy to forget the Outback was just outside.

at the end of the meal and say 'Thanks for the tucker folks, my laptop beckons, see ya'. However, Neil would have none of my protestations and showed me to my room.

After a few beers, and my first home cooked meal in ages, I was very pleased about acquiescing to Neil's offer. As I sipped another cold beer and soaked up the same TV programs that the rest of Australia was watching, it was easy to forget the Outback was just outside.

Delivering TV signals was one of the jobs of remote shires like Murchison. Some would argue that a remote shire's most important responsibility was ensuring that the dirt road networks remain in a navigable form. However, the more indolent may argue the most important function is maintaining uninterrupted television transmissions. This weighty responsibility involved sucking Australian TV transmissions off a satellite, bunging the signals through a booster box and rebroadcasting them to the locals, which in Murchison's case equated to the 30 or so people who were within striking distance of the low power transmitter. The television, combined with the beer, helped bypass my work anxiety. I forgot about the laptop and settled into an evening in front of the box, chatting with Neal and Elsie.

I awoke the next morning to the comforts of more home cooking. The food was reason in itself to find an excuse to hang around longer, but I was conscious of not wanting to stretch their hospitality. Also, ahead of me was a 400-k day traversing roads of an unknown quantity, so I was keen to get going. Neal assured me that the next 165 k's of road, the distance to the Murchison border, was in great condition, but what lay beyond there was a bit of a lottery. I got the road briefing while Neil gave me a guided tour of the Shire HQ. The tour started in the works depot, which was full of big trucks, tractors and graders. I was in Tonka heaven and quickly developing a soft spot for the Shire. As I met some of the works crew I began to realise where Neal and Elsie got their enthusiasm from—everyone was very proud of their townless patch. In fact they were so

L to R: Murchison highlights – 'The Shire with no town'; Neil Warne (shire boss and quick to offer a beer); bloody big spiders and a shopping list.

proud that the Shire bagged Australian Community of the Year in 1998. I was not sure what the criteria was for gaining such an award but if it had anything to do with road building, more gongs would soon be lining the walls of the shire office.

Murchison took its road building very seriously—being only a few hundred k's south of the Tropic of Capricorn they had to. Their location meant that when the occasional cyclone battered the west coast, huge deluges of rain could end up being dumped on Murchison. Inundations, though, were rare; the shire was in its third year of drought and on target for an annual rainfall figure of just 4 inches. Nonetheless, the rains do come and would quickly close 1721 k's of the shire's 1723.2 k's of road network. (the 2.2 k's of road that remained open was all weather tar. However this probably wasn't a lot of use as the 2.2 k's was made up of many small strips of tar placed before and after the shire's countless cattle grids. This detail came to light when I was dryly told the shire was undertaking a tar expansion project and aiming for 2.8 kilometres of black top within the next 12 months). Despite Murchison's tar advantage, rain remained an issue for both it and all the surrounding shires. Murchison, however, would recover and be open for business faster than the rest of them, and it

Rather than embark on epic weekly journeys to the nearest store, Murchisonians had the store come to them.

Just in case there were any doubts that I was actually in the bush I thought some snaps of a bloke knocking together a cattle grid and shearers would put any such concerns to rest.

was all thanks to a 'bun'.

'Bun' struck me as an odd description for something that was essentially a massive speed hump. 'Speed hump' is perhaps a little misleading, as it looked like it could be hit flat chat, and just to make sure I put a few to the test by attacking them at warp speed. I was hoping for a little airtime, but sadly they were too gentle to allow me to get any space between my wheels and the dirt. Water, as opposed to vehicles and deranged bikers, was the target of the 'buns' slow down campaign. In undulating country, rain can quickly turn roads into streams, causing erosion and subsequently a lot of grief for road users and repairers. The buns were designed to minimise the damage by interrupting water flow; when water hit them it was deflected along the hump and out the edge of the road. To make sure the water didn't just do a quick sidestep around the bun and continue down the track, the Murchison road crew extended the buns for about 50 metres into the bush. The bun–bush extension ensured that not only was the water prevented from getting back onto the road but also that the surrounding scrub got a good drink. This latter point was important, as roads have done a lot to stuff up the natural flow of water, which explains why often one side of a track will look healthy and the other barren.

Bun building, road construction and maintenance are all expensive pursuits, which was potentially a problem for Murchison. The Shire had barely enough residents to fill an average-size pub. The rate base would have struggled to fund building a decent driveway. Fortunately state and federal governments stepped in to fill the void, in the form of road grants. This largesse was not only of advantage to the people of Murchison but also dopey travellers like me, livestock transport companies, mining companies and, perhaps more importantly, the bloke who ran the local mail and grocery run.

The Shire's lack of a town meant there was also a distinct lack of shops. And as man cannot live on roo steaks alone, food had to be bought from somewhere. Rather than embark on epic weekly journeys to the nearest store, Murchisonians had the store come to them. Tuesday, the day I was there, was also 'fax day', which involved the weekly ritual of faxing off the shopping lists. (Faxing off the lists was a relatively recent luxury as they only got direct dial phones in 1985.) Despite this leap in technology, delivering the orders still required the same vast distances to be covered. Travelling around such a large area is not only time consuming but also costly—for some of the remote stations this ends up adding an extra $3000 to the annual grocery bill. The mid-week grocery service also doubled as a mail run. The mail run was bi-weekly, the second run being on Sundays, a bit of a Sabbath drive that involved the mailman hopping in his ute and doing a quick 1100 k's for the day! I had once done 1200 k's in a day from Charleville to Sydney, slaving over it on a 600cc trail bike, but at least I was on tar all the way. I was astounded that this bloke not only covered this distance on the dirt but also delivered mail as he went.

Fortunately I had no insane distances to cover, just a 400-k ride to get me to a place called Mount Augustus. While the distance was more than manageable, it seemed less and less achievable as my tour of Murchison continued. The morning involved a full tour of the Shire depot, offices and the local museum. (Yep, there might not be so much as a shop but they had not neglected their duty to provide a museum.) By the time the tour had wrapped up it was around 12 o'clock and I was anxious to get on the road. I fuelled up, said my goodbyes, and watched another half an hour slip by.

Seventy k's later, and after many mental calculations of distance versus remaining daylight, I began to accept that there was a very slim chance of covering the 400 k's before dark. If I went flat strap, had good roads, and totally ignored my job of photographing and recording stuff along the way, I might have bagged it. I might also win the lottery. The countless times I had faced such hopeless equations in the past I always made a decision to go for it, and then bitched about my stupidity five hours later when I was dodging roos in the dark. The distance/time calculation, however, lacked one minor variant—I had to stop for an hour and a half and take a bunch of phone calls from radio stations. Tuesdays were my update days; radio stations from around the country would dial me up on the satellite phone and find out what I had been up to. I parked the bike and began taking calls. With each new conversation I watched the tree shadows inch further across the road, my language growing faster and more urgent as the silhouettes grew longer. I hung up the last call and made a unique decision—I decided to turn around and head back. I say 'unique' because in over 60 000 k's of riding and reporting I had never failed to finish a day's ride. Returning to whence I left seemed like a failure. On the ride back the shadows flickered across my visor and made the road surface hard to read. The strobing effect obscured a sandy rut that grabbed at my front wheel and flicked the bike, and as I wrestled it back on to course I felt that for once I made the right decision. A little while later I was sheepishly fronting up to Elsie, hoping she had not got around to changing the sheets on my bed.

      The surprising thing about returning was that it did not seem as big a deal as I had feared. I endured a bit of mandatory ribbing, but once that was over everyone said I had made the right decision. Their reassurance went a long way—the locals knew better than anyone the risks of travelling on outback roads after dark. The only downside to the decision was that I was another day behind in the work I was supposed to have done, but as I sat down to more home cooking, and another beer, I decided that I would worry about that tomorrow.

Outback title fights

When I left Murchison—for the second time—the sun was sitting low in the east and the trees were casting a cool 9 am shadow across the road. Nine was a good start for me. I don't believe in early starts; they are uncivilised and bad for my hair. My disposition was also partly based in practicality, as preparing to ride each day was a mission. The three panniers and tank bag had to be strategically packed each morning, otherwise everything wouldn't fit, or it would fit by force, but crush and damage the contents in the process, or, if things were badly packed, I ran the risk of stuff rubbing together and holes being worn into expensive electronic equipment.

After the science of 'pannier packing' was dealt with, I would then ferry the gear out to the bike, clip it on and lash it down with extra reinforcing straps. The final step in the launch sequence was to climb into my riding pants,

boots, back protector, jacket, Camelpak, helmet, sunglasses and gloves. If I were looking to make a quick getaway each day I would have bought a car.

With the morning ritual behind me I settled into the ride. The bike flicked across the landscape but my thoughts were still in Murchison, pondering the tight little community and its unique circumstances. I cogitated my way around the museum, the shire office, the depot, the road crew and Neil and Elsie's home. The trip down my shiny new memory lane only came to a halt when I passed the point of the previous day's U-turn and thought, 'Well, no turning back today. I won't be seeing the Murchison crew again for some time and … and … and none of it was on tape!' I realised I had taken a wad of photos but had not interviewed a single person! I was panic-stricken. Why, after being presented the same opportunity twice, had I failed to roll any tape? Could I not see that there was a decent yarn there? Was it because I was so far behind in my work that I did not want another tape added to my growing pile of 'stories yet to be edited'? Did I do that horrible electronic media thing and think no one would make good recording talent? Whatever the reason, the opportunity was now well and truly gone and my gut tightened over screwing up so badly. I resolved to try and compensate for it by writing more for the website, but the good intentions didn't bury the feeling that I had stuffed up.

> If I were looking to make a quick getaway each day I would have bought a car.

If music calms the savage beast then beauty must appease the self-berater. I was entering the Gascoyne, the terrain was beginning to undulate and small gorges were appearing. The Murchison folk had told me to take a couple of minor detours en route and check out Wooramel Gorge and Bilung Pool. Bilung was a beautiful still waterhole, cool at the edges and a mirror to look at. As I trudged back to the bike through dust and sand I appreciated the true definition of an oasis—Bilung was under siege from a hostile landscape at every quarter.

As the day wore on and the k's rolled by, I grew more grateful for my decision to abandon the previous day's attempt. The roads were average, but there was no way I would have made my destination by dusk. If I had committed to the first effort it would have been a race—a frantic afternoon giving it the beans that may have gone horribly pear-shaped.

The early start I'd made with 'attempt two' gave me time to enjoy both the countryside and the riding. The bike was now handling like a dream. Ditching the dud tyre back in Geraldton had improved my lot no end; the bike tracked perfectly, doing exactly what I wanted and more. My confidence was on the up, I was working with the bike as opposed to battling against it, the tension in my riding had gone and I was enjoying every minute of flogging through the landscape. The only issue was trying to keep a lid on it—the faster I went the better the bike felt. Fun wrecking self-preservation eventually kicked in and I accepted that if I wanted to have a remote chance of dealing with 'the unexpected' I would have to come out of warp speed. At a saner pace, I settled into a rhythm of dodging and weaving around the endless obstacles that present themselves on outback roads: a zigzagging dance leading a cloud of billowing dust. The last 90 k's into Mount Augustus were a little more demanding; the road was sandy, rutted and corrugated, but with the new ring of confidence fitted to the back wheel it was fun, as opposed to a trial. After 3000 k's of battling a crap tyre I was enjoying my riding again.

Fifteen k's from my destination a mound began to slowly grow from the distant landscape. The closer I drew, the more the lump bulged; the earth was developing a Quasimodo-like affliction. The flat horizon was being sucked up into this growth until eventually all I could see was the hump. Mount Augustus obscured all else. It seemed fitting someone had bestowed on it the title

> If music calms the savage beast then beauty must appease the self-berater.

> The only issue was trying to keep a lid on it—the faster I went the better the bike felt.

Mount Augustus, the world's biggest rock, set to a backdrop of lousy weather that was about to cause me a whole bunch of headaches.

of 'world's biggest rock'. I always thought this designation went to Uluru, but it turns out that, like boxing, rocks too have title divisions. Uluru holds the belt in the monolith division. Mount Augustus, weighing in at approximately twice the size of Uluru, is king of the monocline weight division. The locals told me the difference between a monocline and a monolith is that one has vegetation on it and the other one doesn't. I was a little dubious about this. After all, Uluru did have a few stray bushes and trees growing on it, but certainly nothing to the extent of Mount Augustus. Whatever the technicalities, it was comforting to be able to think, 'When it comes to bloody big rocks, outback Australia creams the rest of the planet'.

> Whatever the technicalities, it was comforting to be able to think, 'When it comes to bloody big rocks, outback Australia creams the rest of the planet'.

The problem with the title of 'world's biggest rock' was that I expected something more stunning to look at than Uluru. (Considering I had never seen a postcard or tourist brochure featuring Mt Augustus, perhaps I should have twigged that this was because Mount Augustus wasn't the 'pretty boy' of the rock kingdom.) Uluru explodes from the landscape in an abrupt manner. Mount Augustus,

however, is a little less hasty and arcs up from the surface like a whale coming up for air. The gods had also been stingy on Augustus' colour scheme; with a surface smattered in vegetation it was ruled out of having an Uluru velvety red radiance at sunset. To be honest, the day was so gloomily overcast it was hard to imagine the rock had any radiance at all. The clouds that had been dogging me for a couple of weeks looked thoroughly fattened and ready to burst their billows. As I rode the final few k's to my camp for the night, rain drops began to slide across my visor.

Rain was definitely not on my wish list—it would turn the dirt roads to slush and I would be stranded for days. Depressingly, the drops appeared to be more and more determined to turn the roads into quagmires. I pulled into the camping ground convinced the weather was hell-bent on making it my home for an extended stretch.

The several tracks leading out from the camping ground all showed recent signs of rain—one even had a large yellow sign across it saying 'road closed'. Unfortunately this was the Pingandy track, the route I had planned to head up on into the Pilbara. I stared at it furiously and thought bitterly, 'First rain and now this, it's a disaster'. It was a disaster because I would have to make a 400-k detour to get around the problem. In fact it was probably at least 800 extra k's, as the eastern detour didn't have fuel available along it. I would now be forced to head out west and look for bitumen. If I was in kick-back traveller mode I would have made friends with the bar for a few days and waited for things to dry out. Unfortunately, as was always the case, time was of the essence. Not just because of the dwindling budget but also because I had arranged to travel on one of the world's longest trains between Port Hedland to Newman. If I wasn't there on time I would miss the ride. I had missed out the previous year and there was no way I was going for the wooden spoon 2 years running.

One man road crews. Jimmy had just driven 600 k's to start grading roads. Hitched to the back of his grader was everything he needed for a few weeks of continuous roadwork … including an emergency stash of six weeks of food and water in case he ever got stranded out bush by a passing cyclone.

I continued to stare at the sign and curse the bloody outback for its fickle indifference. My fuming only eased when I was distracted by a one-man road crew pulling into camp. The grader, which was towing two trailers and a 4WD, did an enormous U-turn. The driver was circling the grounds looking for a place to settle for the night. It's not every day you see a grader pulling a fuel tanker, a caravan and a 4WD, so I did the touristy thing and pulled out the camera and took a couple of snaps. By the time I had fired off a couple of shots I was wondering if I had done something wrong, as my photography coincided with the grader train coming to a halt and the driver climbing down and walking towards me, purposefully and directly.

The bloke heading towards me looked a no-nonsense kind of fella. He was clad in a blue work shirt and blue shorts, with the extremes decked out in Blunnies at one end and a shock of grey hair at the other. When he was a few feet away his hand shot out and he announced 'G'day. I'm Jimmy, how you going?'

I reached for his hand, replying, 'Not bad. However I reckon I would be a whole lot happier if this rain buggered off.'

'I don't think there's much in it', Jimmy assured me.

Trying not to sound like a total selfish bastard I threw in a bit of conciliation with my moaning: 'I know the cockies need it but I kind of need the roads to stay open'.

> I continued to stare at the sign and curse the bloody outback for its fickle indifference.

'I think you'll be right. Where you heading?' asked Jimmy.

Pointing over his shoulder, I said, 'Up that track there, the one with the 'road closed' sign on it. Do you know anything about it?'

'Not much mate. I know it gets really tight and windy up the top. You best speak to the bloke who runs the station. He'll set you straight' Jimmy advised.

I thanked him for his tip and said that I had better go and sort my digs out for the night. I promised that once I got myself settled I would pop back for a beer and a decent chat.

A few hours later I was banging on his caravan door, a six-pack in one hand and a recorder in the other. I figured the life of a one-man road crew might be an interesting yarn, and if it wasn't, then there were a few ales to enjoy.

I needn't have worried about the beer fall-back position. Jimmy had been driving graders and trucks for almost 35 years and had plenty of chat in him. The Pilbara was now his new home; he had moved to Gascoyne Junction to enjoy some fresh air and feel like he was his own boss. There were, however, some 'swings' to the 'roundabouts'.

'Temperatures outside can get up into the high 50s. In summer we've had some of the highest temperatures recorded anywhere in Australia. We've even beat Marble Bar at times and that's the hottest spot in Australia', said Jimmy.

I couldn't tell from his tone if he was saying this as a proud fact or a sad reality. Whatever the intonation was meant to relay, the Marble Bar comparison seemed fair. After all, it was only about 500 k's away.

'You get used to it. I do. I love the heat so it doesn't worry me, but with the air conditioning you do get soft', Jimmy added.

I was thinking the opposite—in 50-degree heat, without air conditioning, softness may have come from organs melting. I turned my mind from liquefied kidneys and asked Jimmy if it got lonely being out here by himself.

'Nah, it's not lonely…well it is lonely in a way, but with the HF radio we have contact with the other graders, so we can chat to them.'

Jimmy said it like they were just down at the end of the road, which I guess they were, except the road was 400 k's long.

'Now and then we get a tourist that comes past who'll have a flat tyre or whatever, and you'll have a chat to them, or you drop into a station and have a chat to them. So it's not that remote. It's not really that bad.'

'Is that also saying something about the outback, that better roads and communications are making it a lot smaller?' I queried.

> 'The outback is smaller—we see a lot more tourists now than we did a few years ago.'

'The outback is smaller—we see a lot more tourists now than we did a few years ago. I would probably count 3 or 4 a day and I might speak to 3 a week. With more modern motor cars and caravans and campervans it's just opened the outback right up', Jimmy replied.

'But that's also made possible by you, with the job you do?' I prodded.

'It is part of our job. We try and keep the roads as good as we can for the tourists, and the pastoralists. They've got to get their cattle to markets so we've got to keep the roads up to standard. Also, we are in a priority area—the shire has a mail contract and the roads have to be opened up ASAP so that the mail can get through', Jimmy said.

It seemed appropriate that the conversation had swung back to the weather, as it was playing on both my mind and the roof of the van.

'Because of the rain, I'm worried that I might not get out tomorrow. If a big rain comes, you might not get out for a while as well?' I half stated, half asked.

'That's true, that can happen, we just sit it out. Simple, we just sit it out.' Jimmy said this in a manner that led me to think he had long ago learnt to live with the foibles of weather and its effect on the outback.

'So what happens during the cyclone season—do you have to carry extra reserves?' I asked.

'Yep, I've got enough food here to last me 6 weeks. If I was stranded I could last 6 weeks. If it's bad they will try and get me home before the cyclone hits or they will try and fly me out afterwards, but bear in mind that this country dries out very, very quickly, so it might only be 3 days between the rain stopping and starting work again.'

'Mate, I hope it's not 3 days, otherwise I am going to be running really late', I joked.

Jimmy laughed and I thought that was probably a good point to flick the recorder off and focus on the remaining beers.

Found on the side of a dirt road, a long way from anywhere or anyone, or to be more exact S25°09.027' E116°55.287'. I defy anyone to find a lonelier phone box.

A short while later I bade him farewell and stepped out into the spitting night. As I walked through drizzle I mulled over Jimmy's claim that one of the reasons he had moved out here was to slow down. His concept of 'slowing down and living a laid-back life' involved working 11-hour days: 10 days on, 4 days off. If that was the 'easy life' I dared not try and conjure up the sort of hours he must have been working when he was driving semis. The hours and the remoteness were offset, however, by some financial incentives. Jimmy said this was a zone B tax region, which meant a $1200 rebate each year. On top of that, the shire laid on an air-conditioned house for the princely sum of $35 a week. I stared up into the sky and cursed the rain; maybe those financial legs-ups were needed not only to offset high fuel and grocery prices but also to compensate for frustrations—like being stranded when rain turned the dirt roads into slush.

The rain was still making pit-a-pat noise as I put my head on the pillow, a sour reminder of the hassles it was going to cause. Despite my frustration, the staccato tap-tap-taps were also comforting—having been out in deserts, it was a sound I had not heard for some time and it quickly nursed me to sleep.

18

'I wouldn't trust the maps out here, mate'

'I've only been up part of the road you're going on, but I know it gets pretty tight and windy up the top.' Jimmy's words rattled around in my head as the bike skipped through another rocky corner.

'Tight! Shame he didn't mention slippery as well!', I fumed as I entered the next corner. At least this one had the camber going the right way, even if the rocks had been usurped by mud.

To be fair, Jimmy wasn't to blame over me being engaged in one of the hardest rides of my life—I had to accept responsibility for being out there.

However, in the midst of the struggle I was still keen to find someone to blame. I re-focused my frustration away from Jimmy and towards the station owner who had told me, 'Nah, the road's only closed to 4WDs. We were going to re-open it today, but there was a smattering of rain last night. You'll be fine on a bike. Just ride around the 'road closed' sign and keep going.'

As much as I was cursing the cocky, it probably wasn't fair to target him either, because he would have thought I was on a light trail bike. If he had actually laid eyes on my two-wheeled truck he might have offered different advice.

I had gone seeking info about the track after waking to silence in the morning; the splattering sounds of rain had vanished and I was keen to find out the state of play with the roads. The first person I turned to was Jimmy, but he said he didn't maintain the track. Ominously, he wasn't sure if anybody ever did, and stuck with his earlier suggestion that the best person to speak to was the local station owner. The cocky initially spoke from a landowner's perspective, expressing disappointment about the rain.

> **It didn't look deep, but judging by the way I had been sliding all over the track leading up to it, riding through it would be like jelly wrestling with an eel.**

'There was bugger-all in it, but I reckon it's going to piss down in the next day or so.'

With the weather report out of the way he went on to say I should be able to get through on the Pingandy track. With the travel health card reading 'Track OK, Weather in Decline' I thought I should get up the track while I still had the chance. Unfortunately this decision blew away my plans of hanging around Mt Augustus for a day. I would have to come back another time to properly suss out the world's biggest rock.

After frantically packing, refuelling and stocking up with junk food, I was on the track. Three k's later I was bogged. The track had disappeared under a large sheet of water. It didn't look deep, but judging by the way I had been sliding all over the track leading up to it, riding through it would be like jelly wrestling with an eel. Water crossings usually meant there was fun to be had, but the state of the track suggested I had a good chance of going a gutser and dunking my panniers (read computer, etc.) in the water. Playing safe was the best option, and for some stupid reason I deemed this should involve riding along the absolute edge of the track. I had wanted to get off the track entirely but it was hemmed in by scrub.

Going on the verge was a great decision—as soon as I hit the water the back wheel began sinking, fast. Within seconds I was inching forward at the same rate as the back was digging down. With the knowledge that some movement is better than none, I abandoned ship and began pushing. A metre later the battle was over: 'Mud' one, 'Drew and Bike' nil. Beaten and breathless I scowled at the mud hole, although my despondency eased a little when I spotted wheel marks on the other side that suggested I wasn't the bog's first victim. With the bike wedged deeply in the furrow there was no need for a side-stand, a truck would have been needed to topple the bike free of the mud's embrace. A winch was an optional extra that BMW didn't provide, so the only way I was going to be able to wrestle the bike free from the quagmire was to lighten the load.

> Vertebrae crunched, veins popped, I sank deeper into the mud and the bike stayed put.

Twenty minutes later I had finished ferrying the luggage through the knee-deep mud. As I wallowed back to the bike for the 'break free' attempt I contemplated what I would do if I failed. Being pig-headed and stubborn, my alternatives boiled down to not failing. No one had helped me out of a jam before and I wasn't going to start now. I had been in much worse binds. Once I had broken down near the Iranian border at dusk. I was in Kurdish country, not the safest part of the planet, which perhaps also accounted for the distinct lack of roadside automotive assistance. I worked frantically on the bike for hours, alternating between chocking my jocks and gagging on the pen torch I had stuffed in my gob so I could see what the hell I was doing. No one came to my aid then and I was certainly not going to let a fellow countryman catch me helpless in a pissy bog. With my machismo dangerously over-inflated I decided I was now in a position to be able to clean and jerk the back wheel out of the bog. Vertebrae crunched, veins popped, I sank deeper into the mud and the bike stayed put.

The next option was to stuff some branches under the back wheel. Hopefully the tread would bite into them and solve the problem. Twenty minutes later, with enough foliage shoved under the bike to start my own composting business, I decided it was time to give it a go. Rather than burden the bike with the weight of my own carcass, I stood next to it, started it, dropped it into gear and let the clutch out. The tyre bit into the branches with an enthusiasm I had not counted on. The bike launched out of the hole faster than I could extract my feet from the mud, and now it was me that was being hauled from the quagmire. With a stubborn conviction to preserve the bike's new found un-bogged status I kept the beans on. The bike responded to my stupidity by taking me for a short ski through the mud. After 10 metres of 'slaloming', we were back on terra firma.

The bog incident had left my bottom half encrusted in mud and my top half splashed and daubed with watery sludge. The panniers were also similarly smeared after I wrestled them back onto the bike. I sincerely hoped the day was not going to be a long litany of bogs, unloading and loading. That was time consuming, energy sapping and I had a long way to go. I saddled back up and eyed the road ahead, grateful to see it winding up into the ranges. Hopefully there would be no more bogs when I got up into the hills. I knew the hills went by the title of the Kenneth Range, but I wondered also if it was the start of the Pilbara. Whatever the title, it looked like there was a wicked ride ahead.

The funny thing about ranges is that they attract rain. I had forgotten this little bit of Year 8 geography when I earlier accepted the station owner's assurance of 'You'll be right when you get up into the ranges mate. The water is only sitting around down here on the flats'.

The further I got into the ranges the more convinced I became that the 'pitter-patter' of rain I heard the night before must have been a deafening roar up in the hills. For four hours I wrestled and laboured along a greasy track, working my way through a million gear changes in the process. Sadly it was a constant shuffle between only the bottom three gears. The gear lever dance started at the bottom of each hill with 1st gear, a flick of the wrist and then a quick jump to 2nd, and if the furrows and rock outcrops weren't too bad on the way up then I

might get to sample the delights of 3rd gear before I ran out of hill. Each hill crest meant the short-lived freedom of 3rd would be quickly replaced by 2nd, usually accompanied by an expletive and an earnest fear of not being able to wipe off enough speed and avoid being smeared all over the bottom of the valley I was looking down at. As the bike lacked a 'deploy parachute button' (despite my frequent re-checking) I would again stab the gear lever down, hunting for 1st and praying for the divine intervention of 'engine braking' to come to my salvation.

To complicate things further, the scrub made it difficult to predict the direction the track would veer when it got to the bottom—usually it was anything but straight. When I got to the bottom, the small valleys were invariably filled with water and lined with claggy slippery red mud. It was a fait accompli that after the downhill Kamikaze runs the bike would never be in the right position to attack these bogs, so the front would slide one way and the back the other. If a front-end slide was survived then the next battle would be fought at the rear of the bike. Fingers of mud clutched at the rear wheel, clamouring to pull it to the bottom of the bog. The bike could slog its way through the bogs if I slipped the clutch, but if I got it wrong the wheel would spin wildly and quickly dig its own grave. Hour after hour of repeated clutch slipping is not recommended in the owner's manual, and the smell of burning clutch was a constant reminder of the abuse I was meting out to the bike. Every burnt waft stirred the thought that if I cooked it I would be truly stuffed—it was remote country and I had not seen another person since I had left Mt Augustus.

If it was any consolation to the bike, my body was also dealing with a similar barrage of abuse: I was soaked in sweat and every muscle ached. All day I had been in 'battle position', standing on the foot-pegs. From here I could swiftly shift my weight about and wrestle the bike through tight conditions quickly, easily and supposedly safely. The 'easy' part was debatable; with such a heavy load I had to be far more physical than normal to get the bike to respond. The quick part was also under a cloud of questions.

The distance on the map seemed to be drastically out— the bike trip meter was telling me I had already done 150 k's, the map was telling me the track had finished 40 k's ago. I suspected the real problem lay with me

My general theory is if you are lost, and you come across termite mounds like this, you can at least be reasonably confident you are north of the Tropic…Which is probably about as useful as saying 'If you come across termite mounds you know you aren't in Antarctica.'

continuously referring to the map, my philosophy being that looking at it more than once a day was bad karma. To confuse things more I was also referring to the GPS, which was telling me I was on a track but was refusing to divulge any further details. Usually I never looked at the GPS—it took the fun out of things by telling me exactly where I was all the time. (The main reason the GPS was strapped to the bike was so each day's travel route could be recorded. The routes were posted up on the website and could be downloaded and used to plot where I had been.) However, when things went pear-shaped, like this day, the GPS was kind of useful. I would be able to look at it and know exactly what sort of mess I was in and how distressed I should be about it. The GPS, though, was only as good as the maps loaded into it, and the scant details on the screen suggested the cartographer was just as confused about this corner of Australia as I was. Despite the lack of detail, it was still nice to have the GPS, which gave me assurance about the direction I was heading and helped me resist a few rash impulses to turn around and check if I had missed a turn-off.

The other nice thing about the GPS was that at least I could phone in my latitude and longitude to the rescue services. Mind you, I wasn't sure if they were going to be all that chuffed about flying out to deliver a jerry can of petrol. The confusion over distances had an unfortunate side-effect—I was now facing a

Retina chaos courtesy of the Pilbara—fields of dazzling white quartz in the midday sun.

bit of a fuel crisis. I'd calculated I had enough to make it all the way to Mount Tom Price, but the unexpected extra k's, combined with the heavy going, had chucked those calculations, and my safety margin, out the door. I was starting to think I was not going to have enough fuel to get me to Tom Price and there was the minor inconvenience of there not being any petrol stations before I got there. Fortunately I did not have a lot of time to think about the implications of running out of fuel, because I was too consumed with surviving the track. However, when there was an easy stretch the fuel re-calculations would rage in my head against a backdrop of questions about whether I was still on the right track.

The navigation and fuel concerns put a dampener on what was otherwise a brilliant ride. Sure it was ridiculously demanding, but I was having a ball. It was the sort of riding I dream about. My only regret was not being able to ride a little bit harder; but the self-preservation instinct is quite strong when you are by yourself in the middle of bloody nowhere (especially if you think you are lost).

After hours of battling, the terrain suddenly flattened out and the road transmogrified into a sandy two-wheel track. A few minutes later I was at the intersection I had expected to see about 50 k's earlier. To celebrate being resynchronised with the map (Drew-speak for 'not lost'), I stopped to take a short breather and also pick off some of the mud clagged on the bike. While chiselling off chunks of 'Pilbara', I frantically scoffed a bag of lollies and a couple of bits of fruit (pre-masticated inside my luggage, thanks to the rigours of the track). Being part-pulped was a help, as speed-eating was essential—it was already 2 o'clock and I had another 300 k's to go. Worst of all, I only had half a tank left.

The road gods rewarded me for surviving the Pingandy test by rolling out a lush dirt track built for low-level flying, so I was going to be able to make up for some lost time. The new road punched out through vast fields of quartz, with the carpet of small stones screaming a dazzling white in the afternoon sun. I worked my way up through the gears and settled in at 'battlespeed'. I was now in a two-way race against both the sun and my dwindling fuel supply. It was blind optimism. I knew the 'Benzene Sprint' was a race I couldn't win and that I would have to swallow my pride and call into a cattle station and see if I could buy some fuel. I chided myself about failing—I viewed fuel stuff-ups as the domain of dickheads, and now I was a card-carrying member.

The shadows were growing long by the time I found a station to pull into, and the vacated look of the property suggested a bigger darkness had already befallen it, and that the residents had long ago upped stumps and bolted. The land around the houses and work sheds had undergone a vegetation bypass; the buildings sat naked and unshaded on scorched barren earth. I rode up to the main building and knocked on a few doors. The house appeared to be abandoned, dust covered everything. As I waited in hope of someone answering my knocks I gazed up the hill behind the house and saw it was strewn with vehicles. Most of them appeared to be defunct but some also gave the appearance of still being functional. I wondered what cataclysmic event had occurred to force the residents to pull the plug and leave. I knocked on a door for the third time, the victim in me starting to think maybe someone was around but was lying low. Either they were wisely trying to avoid dealing with a bald-headed bike rider standing on their doorstep or they had just got sick of outback travellers continually pulling in and seeking assistance.

It was ill of me to think such thoughts, but the abandoned/functional status of the property was making me uneasy. I suspected I could probably find some fuel if I poked around a bit, but the place was starting to feel like a sinister setting for a B-grade movie. I didn't fancy explaining to someone at the other end of a 12-gauge shotgun why I was syphoning fuel from their rusty ute. I took the manly option and decided to keep riding. If my map had recovered from its early brain fade I only had another 50 k's to go before I would bump into a

The future being replaced by the past. After three years of unreliable service, 12 grand's worth of solar powered stock water system was being replaced with 5 grand's worth of good old-fashioned windmill.

bitumen road. Once I was back on tar I thought I would probably have a chance of flagging down a passing vehicle and scoring some fuel.

I left the station wishing I could ride with crossed fingers. The lack of life in the fuel gauge was suggesting I had more chance of winning the lottery than covering the next 50 k's. I pulled onto the main track and rode down through an enormous dry river crossing. The banks into it were high and steep, suggesting that huge flows of water occasionally passed through. I had crossed the Tropic of Capricorn a few k's earlier and was expecting to start seeing some gradual changes in the landscape, but having the start of 'wet season' country marked by such a bloody enormous river seemed a little beyond cliché.

Not far beyond the river the mystery of the station was solved—the owners had gone out to do a bit of windmill erecting. I found mother, father and daughter surrounded by winches, pipes and bags of cement. From their appearance I suspected that not only did they 'windmill build' together but they also shopped together—all were wearing broad-brimmed hats, long sleeve shirts and well-soiled jeans. I was very pleased about the 'soiled' bit as I was caked in dry mud and didn't want them to feel like the odd ones out.

The three legs of the new windmill had been freshly concreted into the ground. While the cement was setting, the tower was anchored in place by long ropes, each tied to a ute out in the scrub. A few feet from the new structure stood a couple of weather-beaten solar panels, both craning northwards. The panels were part of what was supposedly a new era in watering stock: the solar system provided electricity for a subterranean pump that sucked up water for the livestock. Andrew, the head of the family, said he had forked out 12 grand for the system 3 years earlier, thinking it heralded the end of the windmill era, but the new technology ended up being a false prophet. Andrew said it was frequently breaking down and he was putting his animals at risk. The answer was to turn to something tried and proven (and also $7000 cheaper).

This should have read "Main' road open to all traffic'. Unfortunately for me I found the sidetracks to be a whole different story.

The 'wind versus solar' chat gave me a brief chance to get to know the occupants of the 'not abandoned' station. I felt guilty about my earlier assessment of the property and put my perceptions down to my brain being fried from the day's ride. The conversation moved on from pumps to roads, specifically the difficulties I'd had with both the conditions and the distances on the maps.

Andrew's response was, 'I wouldn't trust the maps out here, mate'.

I agreed and seized my chance to comment that the error, and muddy terrain, had blown my fuel calculations out of the water. Andrew and family nodded sagely, I was telling them nothing new; such trials were part of their daily life.

> I poured the fuel into my tank and breathed in the fumes of salvation.

I went on to sheepishly admit I was in a bind and that I had earlier dropped into their property to see if I could buy a few litres of fuel from them. They seemed sympathetic towards my plight and pulled a jerry can from the back of a ute. I poured the fuel into my tank and breathed in the fumes of salvation. As the fuel slopped in I lamented to them that I was probably not the first passer-by they had ever helped out. They laughed a bit and said certainly not, but went on to say it was only during the dry because there was no traffic in the wet, the station was cut off. I closed the tank and shoved some cash into Andrew's hand. He was reluctant to accept it but I insisted—he had saved my bacon and paying for it was the decent thing to do. I thanked them again for their help and apologised for not stopping around, but it was close to dusk and I wanted to get to Tom Price before nightfall. They understood, saying they would not want to be on a bike in these parts after dark. Those were, perhaps, not the parting words of support I wanted to hear but I gave them a wave and rode off.

Forty k's later and my wheels parted company with the dirt and kissed the tarmac. I stopped briefly at the T-junction and gazed back along the dirt track; a sign declared it open to all traffic. I debated that! The last leg of my race to Tom Price saw me beaten by darkness, and as runner-up I was awarded thunder, lightening and a soddening deluge. When I finally found digs for the night, tiredness was slurring my speech and taunting my muscles. Everything ached. I unloaded the bike in the downpour and despite being wet, sore, tired and losing to darkness, I was glad I had pushed through. If I had stopped another day in Mount Augustus I suspect the rain would have trapped me there for a week.

Thirty three thousand litres of diesel

It took a couple of sleeps to recover from the Mt Augustus marathon. The ride had strained muscles I didn't know existed. I felt like someone had taken to me with a meat tenderiser. Besides giving me somewhere to recover, staying in Tom Price also gave me a chance to catch up on some work, visit Karijini National Park and watch safety videos. The safety video was part of the deal when applying to travel on the privately owned dirt road short cut between Tom Price and Karratha. The road was the maintenance track for the Hamersley Iron train line. Gaining a permit to travel it required a valid driver's licence and watching a video about how to drive safely

L to R: Just to prove I was at Fortescue Falls in Karajini National Park.

Knox George, same park sans grinning bald git.

The railway service road between Tom Price and Karatha, arguably one of the fastest, most fun, most stunning rides in 50 000 kilometres.

along dirt roads. After 20 minutes I left with a permit and the knowledge that speeding on dirt roads can be dangerous.

It took me only a few hours to cover the track—apparently a short time—and the crew at the ABC station in Karratha suggested I should have paid more attention to the speeding advice in the video. I was definitely guilty, but the track had been the perfect ride of sweeping corners, hills and countless floodways. In my eyes, and I suspect the bike's, the bigger crime would have been not to embrace such a stunning track.

Despite the whiplash tour of the Hamersley Ranges, one thing I was learning about the Pilbara was that it reeked iron ore. The dirt was rust red, the puddles in tracks were burnt orange, the rocks in the ranges screamed out in every obscene shade of crimson. Yep, if you wanted to make a buck, this was the place to start digging, but if you wanted to be serious about it, you had to invest in some bloody big shovels. Considering my obsession with big equipment, this was a dangerous place to be: the scale of the mining operations in the Pilbara had the potential to cause me sensory overload. Working from the basis of 'too much is never enough', I had organised to fully immerse myself in 'big stuff' by helping a bloke shift 48 000 tonnes of iron ore.

Port Hedland was where my two-day crash course in the bulk handling of iron ore would begin. It all sounded easy enough:
1. Hop in a 3.8 k long train and trundle 426 k's south to Newman.
2. Hand the keys over to some other poor bugger who would spend the night loading up the 330 carriages while we skipped off for a beer and a kip.
3. Pick the train up the following morning and drive home.

To be honest, there was a bit more involved than, say, filling the back of a ute up with loam from the local garden centre. In fact it's so complicated that professional train drivers have to undergo 6 months of training in a simulator before they are allowed near one of these beasties. Fortunately they were not sending me down to Newman with a bloke who had just got his P-plates. Geoff, from what I could work out, was pretty much the head train driver.

I had met Geoff when I was travelling through Port Hedland the previous year and since then he had been promoted. I was pleased for him, however it seemed he was now mainly focused on training other drivers and he sheepishly admitted to me that he hadn't driven a train for some time. I tried not to dwell on the

> The dirt was rust red, the puddles in tracks were burnt orange, the rocks in the ranges screamed out in every obscene shade of crimson.

fact that the driving skill of the bloke who was going to be at the helm of this 60 000 tonne snake might be a little on the rusty side.

The ore journey started early when I parked the bike in Geoff's shed—sadly the train was designed for carrying ore only. As I ferried gear from the bike to Geoff's 4WD I mused on whether I was going to suffer from either separation anxiety or jubilation while being apart from the bike for first time in the journey. But after 10 relaxing minutes of being chauffeured around in Geoff's 4WD I had forgotten about the bike. We drove over the bridge that unofficially divides Port Hedland from South Hedland and Geoff pointed down to the railway yards and said:

'There's our train'.

I could see the front of it, but I was buggered if I could make out the back of it. Geoff assured me it did have an end and gestured to some small hills of iron ore in the distance. Apparently the back of the train would be found in a valley of the brown range that hemmed in one side of Port Hedland. The peaks of this range were actually huge stockpiles of ore. Although the range wasn't big enough to create its own climate, it probably warranted being sketched in as a topographical feature on local maps, although the maps would have to be redrawn each time they carted off a mountain and dumped it in a boat.

The mini ore mountains and red iron dust sums up the 'Welcome to Port Hedland' experience. The dirty brown stain that covers the town is conjured out of sight near the gates of the ore handling facility by a lush green perimeter garden that distracts the eye.

Geoff and I passed through security where I was issued a pass and told to keep it on my person for the next 48 hours. I was going to suggest that I guessed it made the bodies easier to identify, but I spotted a large billboard detailing the number of days since the last injury, and for once my mind beat my mouth and I kept my lips shut. Geoff and I drove down to train HQ; he had some paper work to sort out while I was given a quick tour of the building and the train simulator.

I had always thought that driving a train was a fairly straightforward process of stop, go, stop. However, a massive room filled with a train cabin navigating its way along a video-projected track suggested there was way more involved. The Hedland line is renowned for regularly operating the largest trains in the world and also holds the record for the world's longest train—7.3 k's in length and weighing in at a smidgin under a 100 000 tonnes. With such monstrous beasts, both in size and value, the drivers have to be highly skilled. Impressive as this was, all I wanted to do was have a play on the simulator and see how quickly I could crash a train. Fortunately for the simulator people, time was against us, as I was busy videoing the thing for another TV story. Wrangling with a tripod and camera in a confined space quickly saw half an hour evaporate. Geoff was soon standing at the back of the simulator informing me that it was time go.

We piled into the 4WD and began making our way along the train, with Geoff giving me a run-down of what we were looking at.

'There are six locomotives: 2 at the front, then 110 ore cars and 1.2 k's back you have another 2 engines, then another 110 cars, then 2 more locos, then after them another 110 ore cars. All the locos are worked by remote control using radio signals from the lead locomotive.'

If we hadn't been driving along the length of the train, Geoff's description would have just sounded like a fantastic set of numbers, but ore car after ore car hammered some meaning into his words. As my eyes strained to find the front of the train, Geoff continued.

'It's quite difficult to get them to work the way you want them to work because it is so long and because of the train going up and down hills. You might be going uphill at the front, downhill in the middle and uphill at the back.'

I tried to process what he was saying and feebly offered:

'So you might have some parts of the train accelerating while others are braking …' I trailed off, largely because I thought I was about to make a fool of myself. It seemed, though, I wasn't as big a clown as I feared, as Geoff finished off my line of thought.

Impressive as this was, all I wanted to do was have a play on the simulator and see how quickly I could crash a train.

161

'Yep, you can have some breaking, some accelerating, to keep the train nice and tight so that it doesn't stretch too much and break. If you break a train in half it's a lot of work to put it back together. It can take 4 to 6 hours to put it back together, and in that heat and out there it's very difficult.'

I was starting to understand why the simulator was so important: this was a complicated sucker to drive. It sounded a little like trying to drive 3 vehicles simultaneously. The other thing that I was starting to figure out was that train driving was not the cushy gig I had imagined, especially when things went wrong, as it was the driver who had to put a broken train back together.

'You do most of it, but they do send some help, either from Newman or Hedland, depending on where you are. You've got to walk your train, and 3.8 k's to the end is a long way if you've got to carry equipment, maybe a sledge hammer and things like that. It's very hard when it's very hot.'

My mind scurried down the wrong path and my mouth blurted out:

'So you might walk for an hour only to find out you should have brought a spanner?'

Geoff was unfazed by my poor comedy. 'Yes, an hour and a half, 2 hours, depending on where you are and how hot it is. It may be rocky terrain. You have a hand-held radio so you can report where you are, just in case you fall over or get bitten by a snake or something like that.'

When the front of the train finally came into view, I was having trouble with the concept that just one bloke piloted the thing. While Geoff struck me as a smart and efficient person, he wasn't of the superman dimensions I strangely imagined would be required to wrangle with this juggernaut. Geoff was medium height and clad in the mandatory work wear of blue short sleeves, shorts and compulsory neon orange safety vest. Perhaps the only thing that stood him out from the crowd was a fairly well maintained moustache. Nope, no superman. Then again, didn't they say the same about that newspaper reporter guy?

When we eventually arrived at the front of the train I set about loading all my

> Fortunately trains are rarely parallel parked, so the varying length didn't make judging the rear bumper an issue.

Geoff took me, 6 locomotives and 330 carriages on a two day journey to pick up 48 000 tonnes of iron ore from Newman. Along the way he let me annoy the hell out of him with a video camera so I could make a little TV yarn for the kids that watched 'Behind the News'.

recording gear up into the cabin. Half a dozen steps and a few crates of gear later I was trying to convince one of Geoff's colleagues that 'Yes, all this "shit" does fit on the bike'. My gear took up much of what seemed to be an unfairly small amount of cabin space, especially considering the overall size of the train.

Geoff negotiated his way through my equipment, climbed into the driver's seat and began reading computer screens and going through a bunch of checks. A short while later radio confirmation came through from mission control—we were clear to go. Geoff set about getting the wheels turning— engines revved, clouds of burnt diesel plumed, electric motors whirred and…and… pretty much nothing happened. Eventually, almost imperceptibly we began to inch forward; a slug in a walking frame would have beaten us off the line.

'We are taking up the slack, then we will be off', Geoff said.

Apparently a fully loaded train stretches and shrinks by up to 80 metres. Fortunately trains are rarely parallel parked, so the varying length didn't make judging the rear bumper an issue. It did, however, mean that the train had to be slowly stretched before the power could be applied in earnest. Amidst the sound of engines screaming, wheels squeaking and metal groaning Geoff somehow felt a clunk.

'Feel that? The last carriage is now on the go', he told me.

I wasn't sure if he was pulling my chain or not, but the way he gunned the engines suggested he was happy that the slack was out of the train. After a lot more engine wailing we were becoming stiff competition for the fittest of molluscs.

We swayed across a series of points and Geoff began to explain a few track-side safety features, equipment that suggested he didn't always know what was happening 200 cars back. Banks of regularly placed sensors monitored wheel temperatures; if there was variation it would mean that a brake was jammed on. When that happened, Mission Control would order the lucky driver to stop the train and go for a walk and fix the problem. Derailment sensors were also part of the safety array; if a car came off the track the driver would be informed they had themselves a 'loose caboose' (how often do you get to use that phrase for real?). Geoff told me a story about a car in the middle of a train that jumped the track but the driver couldn't feel it and continued on his merry way. By the time the problem was detected the car had clipped the ends off a bunch of sleepers and done $8 million worth of damage. (I was trying hard not to let the numbers faze me—the loco we were sitting in was worth $6 million and it didn't even have a decent sound system!)

The engines laboured on and we slowly inched our way towards Newman. Throughout the journey I fluffed around trying to get shots of the countryside and the train snaking along behind us. (There were only two brief instances in the entire journey where the end of the train could be seen—I missed both of them!) It was also a day where I was forced to chill a little, there were only so many things I could video and I didn't have to worry about the driving, which was a nice change. Perhaps the strangest thing of the day was that after a while my jaw ached. It was the first time in ages I had spent a whole day with someone. I liked Geoff's company, he was an affable bloke and we chatted for much of the 9-hour journey.

However, the end of the day did not come without some hiccups—a passing loaded train was having troubles and we had to unhitch a couple of our engines and give it a push up a hill. This simple-sounding act took a few hours of unhitching, pushing and re-hitching. The result was we got into Newman after dark. Geoff apologised for the extended journey but I assured him it had something to do with me as darkness nearly always beats me to a destination. We handed the train over to the night shift, then hopped in a 4WD and bolted to get some grub and a bit of kip.

L to R: This is what happens when you let a fool play with the time lapse function on a camera while they are sitting in a slow moving train at night.

Two hundred and forty tonnes of dump truck with driver (or more correctly 'Production Technician') Caroline and her supervisor.

The next morning involved a whiplash tour of Newman and the mine, Mt Whaleback. I am not sure who bestows titles such as 'mount' on geological features, but if it had been my job, I'd be asking for the title back. The mount was now a divot—an exceptionally large one. I was told proudly that it was the largest open-cut mine in the southern hemisphere. (Mind you, I was pretty sure I had heard the same line when I visited the 'Super Pit' in Kalgoorlie.) Whatever the deal, it was a ridiculously big hole and, thanks to one of Geoff's mates, I was getting a first-hand view of it. The tour of the mine also had an objective: I wanted to meet one of the women who drove the 240 tonne dump trucks around the pit. Unfortunately the best laid plans go astray and the word 'media' had a lot to do with it.

Someone along the mine's chain of command had got the wrong end of the stick; there was a degree of nervousness about my visit and what I was interviewing the woman about. (Which was pretty funny considering all I wanted to know was 'what's it like driving a bloody great big truck around a mine and do you ever drive over cars by accident?'—I was going for a Walkley award with this one.) Caroline, the driver they had selected for me to interview, had been infected

Posing in front of $20 000-worth of tyre, the company chews through 4 million dollars worth of these things each year.

by this nervousness and was very anxious when we met. To make matters worse, I ended up having to do the interview in front of her boss, something that did not relax her any further. I found out later that she had reason to be nervous: she had been advised to be careful about what she said. The confusion stemmed from some industrial issues going on at the mine; someone at a local level had got it into their head that this was the reason I was there and didn't want the issue discussed. The problem with this kind of approach was that it got the opposite result—people were unnecessarily spooked and I just asked a whole lot more questions. Initially I didn't give a stuff about any industrial issues, but after it screwed a nice little story about women driving flaming big trucks I became a whole lot more interested.

Lack of time, and also consideration for Geoff and his mate, forced me to bite my tongue and leave the issue alone. The real downside was that I didn't end up with much of a story about women working in the mine. The most interesting bit was that they were sought after as machine operators because they were not as rough on equipment as blokes. I found it hard to imagine how anyone could be rough on such massive bits of kit. After all, I only came halfway up the wheels when I stood next to one of the trucks! Damage did happen, however, and it was expensive. Trashing a tyre by ripping a big hole in it was a $20 000 stunt. But this was small potatoes: the budget for replacing tyres on the dump trucks was $4 million a year. It was scary to try and fathom the money that was involved in the entire operation.

It was scary to try and fathom the money that was involved in the entire operation.

With the interview wrapped up, Geoff and I went to collect the train, which was now 48 000 tonnes heavier. Geoff did his checks and received permission to launch. He released the brakes, pushed the throttle forward and the engines began to howl as though they were screaming at the wheels, urging them to turn. Eventually they did, and ever so slowly the train began to pick up speed.

A few minutes later Geoff turned to me and said, 'Watch this, we are on a slight slope, gravity will do the rest', and backed off the throttle.

Just in case you've ever wondered how big the fuel tank on a locomotive is. On the 800 k round journey between Pt Hedland and Newman our 6 engines chewed through 33 000 litres of diesel.

Slowly the train began to pick up speed; the hill was doing all the work, which was apparently a great way to save on fuel. Considering I had paid some rude amounts for fuel in remote locations, I was interested in how much the train chewed through.

'It's about 5500 litres, per loco, per round trip', Geoff told me.

My mind slowly mulled over the number, multiplied it by 6 and eventually blurted out:

'So around 33 000 litres! Struth, your cost—the labour—is pretty incidental then?' I hoped this wouldn't make Geoff feel as though he was being underpaid.

'It is. It's nothing compared to the machinery, the value of the track and the iron ore', Geoff agreed.

Geoff was passionate about trains. To work on the ore trains he had moved his family from Melbourne, the opposite end of the country. I wanted to know if the move had turned out to be everything he had expected.

'It's more than I expected. It's the best equipment, a very good company to work for, and it's a challenge every day to bring a train that size back from the mine. The empties are a lot easier, but when it's loaded the weight is enormous and it's quite difficult. But you get an immense pleasure when you arrive at the port and you look back along your train and think, "I got that back here in one piece, I've done what I get paid for". It's a great pleasure.'

A few hours later the pleasure was dissipating. I suspected something was going feral when Geoff started alternating between cursing and pleading with one of the computer read-outs. The computer was saying there was a lack of brake pressure at the back of the train. When you are sitting at the pointy end of 60 000 tonnes on a down hill slope, 'no brakes' is not all that welcome news. Geoff wasn't overly chuffed about the whole scenario either. He pressed a few buttons, cussed some more and then said:

'We are going to have to walk the train and find the problem.'

That was fine by me; I was all for having fully functioning brakes. The bit I was worried about, though, was 'how do you stop to fix the brakes, if the brakes are telling you they are not working?' Geoff threw a bunch of switches, which from what I could work out meant he was basically chucking it into reverse. Because it was a diesel electric he didn't have to worry about such a manoeuvre causing a gearbox to drop out on the track. Nonetheless, I was glad it was his engine he was caning and not mine. After a prolonged period of squealing the whole show came to a halt. The problem then became keeping it there.

Any car driver knows that if you park on a slope you need to put the handbrake on, and it's the same deal with a train. The only difference is that it has to be applied by hand to the first 140 carriages. After half an hour of walking and spinning the large brake wheels on each carriage, we were finally in a position to complete the trek to the end of the train and suss out what the problem was. As we walked along the train I surveyed the endless horizon and mused that despite the massive size of the train, it, and us, amounted to gnats crawling across the massive rump of the outback. My romancing about the grand scheme of things was interrupted by the arrival of the rescue team. They gave us a lift to the end of the train where hammers, spanners and swearing were all vigorously employed to resolve the brake problem. After what was considered to be an appropriate amount of swearing and hammering, Geoff looked a little happier and we climbed back into the 4WD. On the drive back to the loco there was debate over whether this had resolved the problem. I wanted to believe it had—I was very keen about the 'brake' function.

The emergency stop had seen the train come to a stop straddling a hill, the edge of the Chichester range. Geoff reckoned the hill was the cause of the problem: it had interrupted the radio signal from the end of the train. The lack of signal saw the computer default to the 'Game Over – No Brakes' option. Foiled by

Q. How do you keep 60 000 tones of ore and train in place when parked on the side of the hill? A. Run like buggery and apply the hand brake on 140 of the carriages.

the bush! Straddling a hill presented a problem for Geoff, as he had to balance the power and brakes very carefully to ensure the train did not snap in half. Snapping was one problem, another possibility was a large shunt would ripple through the train and the head loco would end up being flicked like the tip of a cracking stockwhip. Several minutes passed as Geoff juggled the brakes and the power across the 6 engines and 330 cars. Eventually the train moved and all reports were that it was in one piece. Unfortunately, though, we scored the complimentary 'ripple effect' prize; a small nudge in the middle of the train was running down through the cars like a building wave. Geoff told me to quickly wedge myself into my seat and brace myself with my feet up on the dash—apparently the impact could send you flying into the windscreen. Rumbling towards us was a clatter of clunks, like giant dominos falling and we were the last one in the chain. Like a slap on the back, the wave passed through and dissipated into the horizon.

'Not that bad, that was only a small one', Geoff said, and the train continued down the hill.

We got to the bottom of the hill, parked and set about releasing the handbrakes. (This was one of those rare times where the operating manual said it was better to drive down the hill with the handbrake still on.) The brake debacle had cost us hours and interrupted traffic all along the line. It was going to be dark by the time we got back to Port Hedland.

We never got into Port Hedland—Mission Control decided Geoff had been at the helm for long enough. A relief driver was sent to meet us 80 k's out from base, so we got to do the final part of the trip home in a 4WD. I hate not completing a journey—my trip from England to Australia finished in a similar inauspicious manner when I blew my bike up 300 k's from home. On that occasion I came home in the front seat of my brother's car with the bike in the trailer. Victory was snatched from my jaws. The failure with the train was nowhere as significant, but I still felt a little disappointed about not getting to see the look of satisfaction on Geoff's face as we rolled into the port.

I had enjoyed getting to know Geoff over a couple of days. It was a refreshing change from the shorter intense periods I often spent with people. As we headed towards the distant lights of Port Hedland I talked to Geoff about his genuine enthusiasm for his job and how it was a rare thing to find. I recounted to him his words from when I met him the year before, 'I came up here to drive the biggest and best trains in the world and I love it'. Geoff said that was indeed the case, but there was more to it than that. He had also come to the Pilbara to escape driving commuter trains in Melbourne. To be accurate, it was to escape the gruesome part of driving trains in Melbourne. Unfortunately Geoff had experienced, on several occasions, people committing suicide in front of trains he was driving. I got the impression the final straw was running over a young couple who were heroin addicts. Apparently they had made a suicide pact to finish their lives by shagging on the railway track and letting a train run over them. Taking your life is one thing, but to involve another innocent person seemed to be the ultimate act of selfishness. I was a little shocked by what Geoff told me; he was such a decent bloke and it seemed so unfair that shit like that had been forced upon him. Geoff said it was OK now, the move had put all that behind them.

'I don't have to worry about that sort of thing out here', he said, nodding towards the vast open expanse.

Twenty-first century swagman

Before iron ore put its rusty paw print on Port Hedland, the little dot in north-western WA was a frontier town. Even during the construction of the ore-handling facilities, shopping precincts and dust suppression sprinklers, the place still behaved like it was out of the Wild West (because Hedland was already pretty far west, construction crews felt obligated to fulfil the 'wild' bit). To get a bit of an idea about what Hedland had been like, I traced down a local businessman who had lived there for most of his life. Arnold witnessed the town go through its mineral-led metamorphosis; a pupal phase that he assured me was accompanied

Appropriately, between a place called the Walkabout Hotel and a truck stop, I found Terry the Hitcher. Terry is a modern day swagman who had been hitching and working around Australia for the past two years.

with plenty of growing pains. Arnold said it was such a rough town that they even had a boxing ring inside the local pub, which enabled disagreements to be resolved in a semi-orderly fashion. Arnold, now in his 70s, spoke fondly of the bar side entertainment and lamented its demise.

'Unfortunately the publican put a stop to it because the punters were watching the fights instead of drinking the beer.'

Today, much of the frontier feel is gone, but on the outskirts of town there is a motel/hotel still clinging to the past and optimistically calling itself the 'Walkabout Hotel'. It seemed appropriate that out the front of a place with a name like that I should run into a bloke who was living proof that the swagman is not dead. Admittedly Terry didn't have a swag, a beard or a billy, and I doubted he had ever done anything dubious with sheep when camped near billabongs. Terry, though, did live his life on the road, hitching from place to place looking for work. Befitting a life on the move he kept things simple, and all his worldly possessions were packed in a small canvas bag at his feet.

'I'm trying to hitch up north to Broome or Kununurra', Terry said, his eyes hidden behind wraparound sunnies that I guess were also out of keeping with the garb of Patterson's swaggie. (I suppose the same could be said about Terry's cargo shorts and well-worn deck shoes, but I don't think being a swaggie ever had anything to do with fashion, so I'll leave it there.)

'So how long have you been here? I saw you out here last night trying to score a lift', I queried like a nosey copper.

'Probably about 15 hours', answered Terry, without any note of concern.

I was surprised to hear this, especially considering that next to the hotel was a large truck stop that constantly had road trains thundering in and out of it. I asked Terry if it usually took so long to score a lift.

'Aw, not usually, sometimes it can be pretty quick, other times it can take a couple of days.'

A couple of days! My mind was reeling. I had done a little bit of hitching in my travels, including bumming rides from Adelaide to Sydney so I could make a radio series about the experience. The longest wait I endured was 5 hours and it did my head in. The thought of being stuck in the one spot for a couple of days made me nauseous. I wondered if sitting on the side of the road for a couple of days fazed him.

'Nah, it's all just experience I think', he replied.

I was rapidly starting to get the impression that Terry didn't sweat the small stuff. I soon understood why.

'You get used to it after a couple of years', Terry said resignedly.

The figure of '2 years' whipped around in my head as though the synapse responsible for carrying the thought had been ripped from its anchor.

I stammered incredulously, 'So you've be travelling around, living out of that bag, for 2 years?'

'Yep', he said simply. The synapse was still lashing around in my head like a storm-damaged power line, arcing and etching '2 years of hitching' onto the walls of my mind.

Trying to regain my composure and look like I met professional hitchers every day I asked, 'So what happens, are you just looking for work?'

'Yeah, just looking for work along the way. You usually don't have much trouble finding it, it's pretty easy'. Terry sounded completely unperturbed.

'Are you finding other guys travelling around Australia, doing the same sort of thing as you?' I asked him.

'Yeah, there's a few, quite a few actually. One of my best mates is doing it as well. I just left him down in Manjimup, southern WA. The poor bugger just cut his thumb off, working on a farm.' Terry half laughed at this. I wasn't quite sure why.

'Did he get it stuck back on?' I asked.

'Nah, he couldn't, it was just mulch. Turned into fertiliser, basically.'

I didn't think Terry was trying to wind me up, or be particularly macho. He just seemed to see things in a fairly straightforward matter and saw no point in adding unnecessary language. I switched back to asking questions about his travelling.

'So how many laps of Oz have you managed to do in 2 years?'

'Nearly 3, and every time has been totally different. I won't go to the same place twice unless it was really good. I found that each state is like being in a different country, just the way the people act towards you. You can walk into one town and people are offering you work and wanting to chat. You know, storeowners are coming out for a chat, people stopping you for a chat. However, in other places people don't want to know you unless you've been there years. It's like you are some sort of alien.'

'Is it addictive, or do you want to stop?' I asked.

'It's very addictive. I can't ever see myself stopping. I guess I might have to one day, maybe', he answered.

Having spent reasonable slabs of my life on the road, I empathised about it being a habit that is hard to kick. The thing that usually makes people go cold turkey is when the cash runs out, but being broke didn't seem to bother Terry.

'Well, like right now I am. I've got $10 to my name. But them's the breaks, hey? I've got my fishing reel in my bag, so I am not going to go hungry. So I'll just keep going until I get another job.'

'So you're not stressed?' I asked, thinking I would be near catatonic if I were in the same position.

'Nah, I'm not stressed. I'm happy mate.'

'Continually being on the move there would be no dole for you either I guess?' I asked.

Terry looked at me flatly. 'Nah, no dole, don't waste me time with it mate. A lot more money to be earned out there than sitting on the dole.'

Petrol pump policies

Hedland to Broome was a blacktop transit—the map was devoid of a dirt road between them so I was confined to the tarred luxury of Highway One. I guess a purist would argue 'if it's on the bitumen then it isn't the outback'. I can understand that, especially considering I have trouble accepting anything next to the coast as being part of the 'outback'. The stretch of Highway One between Hedland and Broome is not only tarred but also basically follows the coast, although you'd have trouble picking it as it is hemmed in by scrub most of the way. The problem is then: how do you classify these areas—remote?, isolated?

A place that finds itself in this quandary is Sandfire Flat roadhouse, nestled on the side of the tar at roughly the halfway point between Hedland and Broome. Sandfire wasn't always in this quandary, it was bona fide outback until

Sandfire Flat boss fella Ken Norton.

someone rolled a strip of tar past the front door and effectively put it into the 'stateless' basket. The tar might have conquered the corrugations but Sandfire is still a 2000 k drive to Perth, and the folk who live there have to generate their electricity, pump up their water from bores and so on, a lot of factors that would suggest it's well and truly in the outback.

Sandfire Flat roadhouse was established in the early 1970s when a bloke known locally as Ed headed out along the Broome to Hedland dirt track with a ute full of 44 gallon drums. Ed stopped halfway along, jacked up the rear of the ute, hooked the back wheel to a pump and started selling petrol to passing traffic. The petrol stop grew and a roadhouse was eventually built. This version of the birth of Sandfire Flat roadhouse was told to me in Port Hedland. The bloke spinning the yarn went on to detail Ed's approach to bush justice.

Legend has it that Ed once did the 3-day drive down to Perth to visit the tax office. He arrived there at 4 pm on a Friday expecting to be able to sort his business out then hop straight back in his car and start driving home. Unfortunately Ed arrived to find the 'closed for business sign up'. A security guard informed him that they shut early on Fridays and he would have to come back at 9 am Monday. Ed's six-day journey to see the taxman turned into a nine-day saga, and when he finally returned to Sandfire he was an unhappy camper. So unhappy that he instituted a new sales policy for petrol. The new service code applied only to government vehicles seeking petrol after 4 pm on Fridays—they would have to wait until 9 am Monday before they would be served.

Fortunately the bike didn't have government plates, so I wasn't too fazed if the policy was still running, but while I was there I thought I should stick my head in for a chat with Ed's son Ken, who now runs Sandfire, and see if there was any validity in the story.

'Well yeah, I don't know about that', Ken answered. He paused for a moment to collect his thoughts, then continued.

'But I suppose there would have been similar situations of that nature. He was a man of principle I suppose, and an eye for an eye. Actually, there's a little saying in the bar, under the portrait of him: "A man who is true to himself will always have enemies, but among the genuine, he will always have friends". That was his basic philosophy on life', Ken said.

Ken spoke with a rolling growling voice that tumbled out through a grey goatee beard, its long squared-off whiskers filled the V on his polo shirt. The beard started to move again and more words rolled out.

'I know one time we had the State Electricity Commission come in and check our power. It didn't come quite up to scratch and the SEC man was going to shut him down, so dad went and turned the generator off. The SEC bloke said, "Well, I can't get petrol", and Dad said, "Well, that's your problem"'. Ken had a chuckle and then continued.

'Dad said, "You go and do your inspection again and I'll start the power plant up".'

Ken was a big bloke and his chest heaved with a laugh before going on.

'The bloke came back to a realistic situation and he gave Dad a 3-month leeway to get the work up to scratch and it worked out good.'

Ed had dragged his family into the bush to start the roadhouse when he was 60. He had been a truck driver all of his life and Ken reckoned that the roadhouse was the last roll of the dice for his dad. It sounded like a tough gamble.

'We just had a shed, a caravan and we were serving the fuel out of 44 gallon drums.'

'So was it pretty much a dirt track running past your door?' I asked.

'Yeah, pretty much. There was a little bit of bitumen out of both Hedland and Broome but dirt all the way through.'

> 'Nowadays, if people don't get served within 30 seconds it's bad service.'

Considering the number of cars, not to mention the occasional bus, pulling into the roadhouse while I was there, it was a little difficult to imagine what things were like pre-bitumen. I suspected, though, that there weren't many tourists.

'Not a hell of a lot', Ed agreed, 'but the people you got up here were, I suppose, the "adventurous type". That's what you would call them. Everyone had more time. It wasn't as fast as it is today. Nowadays, if people don't get served within 30 seconds it's bad service. Whereas in days gone by it was very much laid back.'

Service is the issue with a roadhouse gig, as it's a 24/7 deal. It's more of a way of life than a job. I wondered if Ken's kids were going to follow in his footsteps and keep the dynasty going.

'Yeah, it's a tricky one. I wouldn't make my kids feel obligated to take it on, it's not everyone's cup of tea with the public demand. I know from myself I put the roadhouse in front of the family … and … yeah … it never worked out … 'cause you're kind of married to the job.'

Ken was obviously talking about something that was a little raw. I felt I should jump in and avoid any more long pauses hanging in the air.

'You would have to be married to the job, wouldn't you? I mean you help out with accidents, repairs and people in trouble. The list goes on and it's a job that never stops?'

'Oh no, you can make it work for yourself. What happens in other situations is that most roadhouses are leased. People get in there and lease them for 5 years. They work around the clock, make their nest egg and then get out and do what they really want to do' Ken said.

I wasn't sure if Ken meant that owning a roadhouse was not the way to go, but since Ken was in that position I thought I would let the question slide and thanked him for his time.

When I walked out to the driveway service area things had quietened down and there weren't any customers around. Seizing the lull, and also an open door, the Sandfire pet peacock decided it was a good time to check what the latest was in the roadhouse shop. The bird's shopping spree was short-lived. It was chased out with someone complaining, 'Yesterday it was the poddy calf in here and today it's your turn, what's going on with you animals?'

Sandfire's newest staff member, Kim, was doing the expelling. Kim had moved to Sandfire as part of her search for a more laid-back life. It was a serious sea change, as she had thrown in a 12-year career as a nurse. If Ken thought things were going a little too quickly these days, he had failed to include this opinion in his staff induction speech to Kim.

'People are different up here. They are more friendly, more patient, more willing to come in and sit down for a chat', said Kim, pausing for a moment.

'Down in Perth, especially with nursing, you don't get that. People come in, they are your patients, they are only your patients, then they go home and you don't see them again. Whereas up here, you see a lot of the drivers and they come in every 2 or 3 days and you get to know about their wives, their families. They need a bit of friendly conversation, that's all a lot of guys are looking for, so you sit down and have half an hour talk with them and they are on their way again.'

The Sandfire peacock on its way to do a bit of thieving from the lolly stand.

As Kim was fresh from the city I was interested in whether she was getting to know the locals. I wondered if it was a slow process.

'Yeah, it takes a while to get to know people. But working at Sandfire, I think they think, "If Ken is trusting enough to employ this girl, well she's in, she's got her foot in the door". There's still a few drivers who come in and give me a bit of a look. They're the older ones, we're talking early 60s, so they've seen a fair bit.' Kim took a short breath, held her mouth open for a second or two, as though pondering whether to continue.

'There is a bit of a stigma about girls who work in roadhouses and that gets me down. A lot of drivers who come in think that because you work in a roadhouse you are up for anything and I'm not like that at all', Kim said firmly.

The sudden turn in the conversation caught me a little off guard. I gathered my thoughts and asked Kim if she thought the stigma was based in any reality.

'Yeah, I do. I've met a few girls that have worked here and a few other roadhouses', Kim said, before adding quickly, 'That's not wrong, there's nothing wrong with that. They've got no ties, no family, there's nothing wrong with that. It does get very lonely up here, and such as yourself now, I'm just enjoying having a talk with you, and some girls like to have that little bit extra company. There's nothing wrong with that but I'm not like that. Maybe one day I'll meet someone, but while I am with Mark I won't.'

I sifted through Kim's statement and suspected she was trying to make sure I didn't view her as either prudish or judgemental; the jumbled reference to her boyfriend back in Perth helped cement that. Obviously the 'roadhouse girl' reputation was on her mind, and I wondered if all girls working in roadhouses got tarred with the same brush.

'Yeah, you do, but you put up with it. You would have heard me last night in the bar, with that driver saying "come on, give me a massage, give me a massage" and I kept on saying no and eventually he got the hint.'

I tried to empathise, but for blokes, unless you're a rock star, it is pretty hard to fully appreciate what it's like to be hit on all the time. (Having said that, I seriously doubt there are too many male rock stars that would consider this to be an imposition.) Being harassed was something Kim didn't seek or want, but I doubted it would be a long-term issue in her life. The roadhouse job was just the first step towards living the life in the bush she dreamed of. But while she remained at Sandfire I imagined she might be setting a few records straight when it came to roadhouse girls.

TRUCKER ROY

'Sometimes you'll hear them, if they burst, but if they just peel off and you are not watching at that particular time, you might miss them, but 9 times out of 10 you will know', said Roy with, certainty about the status of the tyres on his road train.

The thing I really wanted to know was why the hell don't truckers go back and pick the peeled truck retread off the road? Then poor buggers like me on a motorbike won't come close to killing themselves when they fly over a ridge and find half a tyre in the middle of the road. I wanted to know why, but I wasn't sure if I wanted to ask—Roy was around 6 foot 4 and probably weighed 120 kilograms, and he had a massive wheel-nut wrench in his hand. I found Roy halfway along his 3-trailer road train, jumping up and down on a long bar, trying to get the wheel that had shredded its tread off. Changing any wheel is a pain in the quoit, but having to change one on a loaded road train looked like as much fun as having a tooth extracted.

'It's not too bad. It'll probably cost you half to three quarters of an hour, from the time you notice to the time you get going again', Roy told me.

It was nowhere near as bad as I thought, but judging by the amount of shredded rubber I had seen on the side of the nation's roads, I was thinking that drivers must have to schedule a lot of tyre changes into their itineraries.

'It varies, sometime you might do 2 or 3 a trip, other times you might do a couple in a day. I've gone the last 4 to 6 weeks without doing any at all. It all comes down to how well it is all maintained', Roy said.

By the sound of it there were the occasional horror trips, so I wanted to know what happened when he ran out of spares.

'Depending on what you've got on, or where you are, each circumstance is different. Some you'll do repairs on, other ones you've got to tie an axle up so you can make it to the next town and get it sorted out, whatever it takes.'

Not my handiwork, in fact I suspect who ever did hit it would have been equally sick.

With no north–south railway line in WA, trucks are a lifeblood. Because of this I wondered if trucking was a little different in the west.

'Hell yeah, the eastern states don't have the road train configuration, much. Road trains really are a lifeline in WA. Over east they are mainly single trailers, express trailers overnight. Like Sydney to Melbourne is only 740 k's. It's an overnight thing.'

Nothing in WA was overnight—the place is just too ridiculously big. The size, and hence the distances, bring with them their own set of problems, fatigue being one of them. WA road legislation doesn't require the use of logbooks; drivers don't have to record their hours or when they last took a break. The theory is that logbooks don't work in the west because of the distances involved. Because of this I assumed fatigue was the biggest risk faced by drivers, however Roy reckoned there were bigger ones.

'Either cattle or tourists. Idiots with caravans—either they don't know how to load them up and their headlights end up pointing in the truck driver's face so they can't see anything or else they don't know what they are doing. They get a truck near them and panic.'

I told Roy that my view on caravans was a little different—they were moving chicanes for me to weave through. Cows, though, were a different matter, and I was surprised that they were also a hassle for trucks. Roy assured me that all truck drivers would try and avoid a large cow because they can roll a truck. He went on to tell a horror story of a mate who hit one and ended up a paraplegic in the process.

On this positive note of death and destruction Roy said he had to make tracks. He climbed into his cab, gave a wave and began to get the truck under way by shuffling through a litany of gear changes. I stood watching him fade into the distance, mulling over the joys of the free-ranging Brahmans that seem to roam all over the roads of northern Australia.

㉒
Broome-time

I love Broome—well, at least I did on my first visit. However 'frame of mind' always colours how you view a place and unfortunately all I could see was my own backside when I pulled into town the second time around.

I was once given a bit of wisdom about getting the most out of 'life on the road' by a fellow traveller I met in Pakistan. His insight was simple: 'travel is work'. At the time I thought this a little ridiculous—after all travel is one of the ultimate self-indulgences. Eventually, though, I understood what he meant; it was about keeping your batteries charged so you can get the most out of what you are experiencing, which requires occasionally taking a break, resting and rejuvenating. I knew the theory but I was caught in a quandary: exploring the country was my job so I didn't really view myself as a bona fide traveller. Consequently I never took

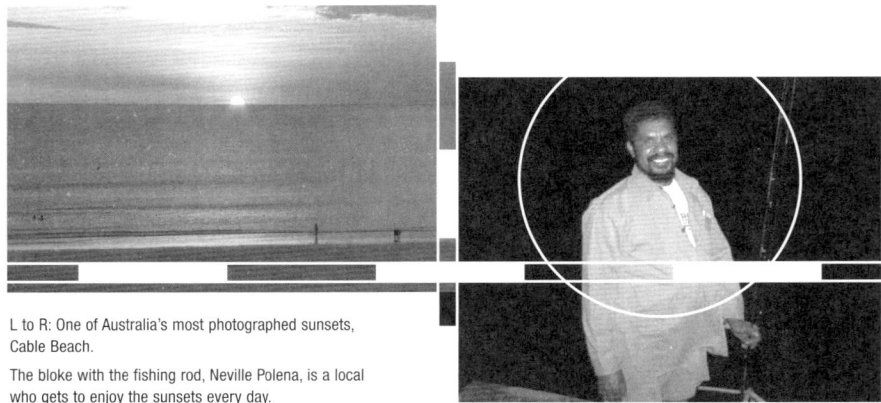

L to R: One of Australia's most photographed sunsets, Cable Beach.

The bloke with the fishing rod, Neville Polena, is a local who gets to enjoy the sunsets every day.

a Saturday, a Sunday or any day off. If I wasn't riding or interviewing, then I would be in front of the laptop churning out the yarns. This was the only way to keep abreast of the workload. Despite the extra hours, I continually felt like the luckiest guy in the country to be doing what I was doing, but by the time I reached Broome I hadn't had a recharge day in months. I was stuffed. This was manageable—until things started going astray back at home base.

I had left town with the assignment of reporting on the non-stereotypical side of the outback. Back at Mission Central this had somehow morphed into reporting on how technology was changing lives in the outback. Confused? Try being in my boots! I guess I shouldn't have been surprised. The project was way too complicated; I was filing for multiple mediums and multiple programs, and at the end of the day I was really the only person who knew what was going where. With so many vested interests, someone was bound to get the wrong end of the stick. I just didn't enjoy being beaten with it.

> Being stressed out of my scone was never going to last in a place like this.

The tiredness, the relentless work, the conflict with Mission Central all turned into one big black vortex by the time I passed the 'Welcome to Broome' sign. I could not sleep, I could not think straight. My mind was trapped on a roundabout of the same thoughts and refused to get off. When isolated on your own, 'loops of thought' like this are hard to break out of. I needed the therapy of a travelling mate, someone to have a decent chat with, someone to encourage me to drink my own body weight in beer, while gently suggesting that I was being an irrational lunatic.

Broome is just not the place to hit a mental black spot and career off the road. The town operates on 'Broome-time'. The place appears to have had a clock bypass—no one rushes, things get done when they get done and being chilled is a way of life. Being stressed out of my scone was never going to last in a place like this. After a while the fog began to lift and I worked out I had several options:
• Sell the bike, buy an o.s. plane ticket on the company credit card and leg it.
• Find a doctor with a firm belief in liberal doses of quality pharmaceuticals.
• Advise those back home to re-read the project proposal.
• Humour those back at Mission Central and get on with it.

All of the options had merit. Selling the bike and legging it was particularly tempting, but unfortunately I didn't have my passport with me. In the end it was something else that helped me move on. At first I thought it was Broome-time, but on reflection I suspect it was hooking up with a born and bred local by the name of Neville.

'We've got beautiful water, lovely beaches and the weather is perfect. It is paradise. You really can't walk around in paradise feeling down all the time or there is definitely going to be something wrong with you', Neville said to me.

Neville practised as he preached. He was always armed with a huge and ready smile that shone an impossible number of teeth. Behind that wall of pearly whites was an air of calmness, of serenity. His voice was melodic, hypnotic, washing away my tension as each word passed over me. Neville also spoke with a familiarity and directness; we were quickly into that zone of comfort usually only found with long-established friends.

I had tracked Neville down to a wharf where he was doing a spot of late night fishing. I had been told he was the bloke to speak to if I wanted to work out what made Broome such a unique place, stuff that was way deeper than the aesthetics.

A long time ago Broome achieved what Australia has been rabbiting on about for the last few decades—a multicultural society. Ironically, Broome achieved this through the introduction of the White Australia Policy. Not everyone celebrated the founding policy of Federation when it was introduced. The pearling industry in particular was not pleased because the new policy

BROOME JAPANESE CEMETERY HISTORY

THE JAPANESE CEMETERY AT BROOME DATES BACK TO THE VERY EARLY PEARLING YEARS AND BEARS WITNESS TO THE CLOSE TIES JAPAN HAS WITH THIS SMALL NORTH WEST TOWN. THE FIRST RECORDED INTERMENT IN THIS CEMETERY IS 1896.

DURING THEIR YEARS OF EMPLOYMENT IN THE INDUSTRY A GREAT MANY MEN LOST THEIR LIVES DUE TO DROWNING OR THE DIVER'S PARALYSIS. A LARGE STONE OBELISK BEARS TESTIMONY TO THOSE LOST IN THE 1908 CYCLONE. IT IS ALSO RECORDED THAT IN THE 1887 AND 1935 CYCLONES EACH CAUSED THE DEATH OF 140 MEN. IN THE YEAR 1914 THE DIVER'S PARALYSIS CLAIMED THE LIVES OF 33 MEN.

THERE ARE 707 GRAVES (919 PEOPLE) WITH MOST OF THEM HAVING HEADSTONES OF COLOURED BEACH ROCKS.

meant they would no longer be allowed to bring in Asian divers to trawl the sea floor for pearls. The pearling industry, which was bringing in good export dollars to the fledgling nation, said the new policy would force them offshore because 'white men can't dive'. The truth of the matter was that Europeans wouldn't put up with the risks, poor living conditions and low wages, but truth has never been allowed to get in the way of a good bit of lobbying. The pearlers won out and were allowed to continue to bring in Asian divers, but it was on the condition that they were male only and single. This 100-year-old decision had some interesting results, and Neville was one of them.

Neville's dad came out from Timor to be a pearl diver. Like many lonely divers before him, he ended up forming a relationship with a local Aboriginal woman and eventually they got married. Neville reckoned that 100 years of stories like this was what made Broome a unique place.

'I think it was the mixing of the cultures of the Asians and the Aboriginals. I believe that our customs and their customs are pretty much the same. In the way that our children come first, after them come the old people. After the old people are taken care of, then it becomes us, the middle people, the adults of today.' Neville trailed off, distracted by a nibble on his line. He gave it a short sharp yank and an empty hook flew out of the water. As he dragged the hook back up for rebaiting he continued the story.

'This town is very strange, everyone in this town gets on so well. One thing about this town is we don't see colour—colour is not the first thing we see on a person. We see if that person has got a chip on their shoulder or whether they have got a smile on their face. That makes a big difference to how you are judged here. Things are changing now because there is a more western society, with European people, Caucasian people coming in here. The trouble is they come in here with totally different ideas that don't work in this little town. They need to come to this town, and live in this town, with the ideas that are already here, not bring something from another town that doesn't work and try and change this town. The influx of western ways means we lose a lot of our own culture, a lot of the eastern ways about looking after each other. In western society we don't look after each other, we look after ourselves, and as more people move here, that is what you will see', Neville said. He finished rebaiting and flicked his wrist to cast the hook back out.

Neville was very proud of his little patch and lamented how tourism had changed the face of the town. He described a harmonious town of his childhood, where everyone knew each other, all doors were open and every house was welcoming. Neville reckoned the western influence was slowly changing this, but even so he was still optimistic that it would always be a harmonious place because of its idyllic location. However, this modern paradise had been born out of a painful system of segregation, a system that Neville's family had first-hand experience of.

'That was the Dog-Tag system. The Dog-Tag was just a name we gave it because it was just like a dog tag that you wore around your neck, like the armed forces wear. It was just an identification number and what happened was you needed to get three or four prominent white people to vouch for you, to say that "yes, you could be civilised". What happened was that people would apply on your behalf to Native Welfare, to say that you could be civilised enough to live in town.' Neville paused to gather his thoughts before continuing.

'As an example, on my mother's papers they wrote quite blatantly—they didn't care what they wrote. The bishop wrote: "I believe Dorothy Watson is a good sort. I believe she could clean my house any day and other services I require". These are things that, if you read below the bottom line, you can see the intentions of what this Bishop was up to. I mean he is not the only guy, a lot of the white people of that time wrote those sorts of things.' There was no malice in Neville's voice, he was simply telling it how it was.

'Anyway this Dog-Tag system, in order for you to get it, and live in town, you also had to denounce your Aboriginality, which meant you couldn't talk to your family or any other person unless they had a Dog-Tag. Outside of Broome there's a community called the 'One Mile Reserve' that used to be also called the 'Common Gate'. Before sundown, all Aboriginal people without a tag had to be past the gate. If they were in town after sundown, they would be incarcerated if they didn't have a Dog-Tag. So what happened is on one side you have the Stolen Generation, which were taken away. Then there is my mob, which is a Lost Generation, because our culture was bred out of us. We try to search for it now, and luckily there are still a few elders alive that we can learn some things from.' Neville paused for a breath, looked down at the water and gave his line a bit of a jiggle.

'When I talk about the Dog-Tag system, I do that in a positive neutral way. It has no ill-feeling from me because I need to learn Western ways in order to survive in this society of today. A lot of my culture of yesterday is no good to today's society, but it is very good to myself as a human being. To know where I came from, how to live off my country and all my Dreamings of my country. I can still do the traditional hunting, but as far as all my Dreamings, I will probably never ever find all that stuff, ever again. A lot of that is now lost, so how can I hand that down to my children?'

㉓
Team Dutch Commodore

To get from Broome to Kununurra I took a short cut called the Gibb River Road. To be honest, only a lunatic would call it a short cut. On paper it's a smidgin shorter but when you factor in car-eating corrugations, shin deep bulldust and countless river crossings, it's the long way to go about things. However, the torture meted out by the Gibb is worth it—it cuts through the heart of the Kimberley and it rates as one of Australia's legendary tracks. Not only is it remote and a hard slog, but it also features gorge after stunning gorge, loads of wildlife and countless ranges.

  I had planned to do the Gibb in my travels the year before, but unfortunately I was forced to do an abridged version because the roadhouse halfway along was shut and there was no way I could cover the 700 k track on one tank. The lack of fuel forced me to do snippets of both ends; from the west I cut

It takes a special type of dill to turn their back on a croc, (freshy or not) and take a self-portrait.

down past Windjana Gorge, Tunnel Creek and then on to Fitzroy Crossing, and from the east I doubled back past Emma Gorge to the Pentacost River. The abbreviated version was still well and truly worth it. Windjana was not only stunning but also packed with freshwater crocodiles. I had been keen to get close to a few 'freshies' after being assured countless times that they were not aggressive, unless of course you are dopey enough to provoke them. While I was at the gorge there were 3 Dutch blokes putting the theory to the test, all swimming and taking turns at getting photos of themselves next to some decent-sized crocs.

The 'me next to a crocodile' photo shoot appealed to me also, so I set about taking a few self-portraits standing next to a sun bathing croc. However, I drew the line at swimming with the beggars—I figured 'why put my head in the lion's mouth?', so to speak. Mind you, it's debatable how smart it is to turn your back on a croc so you can take a self-portrait of yourself next to it. My lapse of logic aside, taking risks seemed to be the raison d'etre for the Dutch boys. The crocs were just another part of their adventure of travelling down the Gibb River Road in a family station wagon, which led me to ask them a few questions, like:

'What the hell are you doing going down a 4WD only track in a 1980s Commodore station wagon?'

'To scare the shit out of all the Aussies,' was the reply.

I laughed and wanted to know if it was working.

'Well we get some funny looks, definitely, but if you take it really easy, it's not a problem. It's just a matter of the bigger rivers, the Pentacost and the Durack. But if you take some Kimberley currency with you, which means sixpacks of beer, then hail down the first 4WD that is going by and say, "Hey mate, how's it going? You want to earn a sixpack, towing me through?"' The Dutchman laughed before continuing:

'So we just wrapped a big tarp in front of the radiator, put some cloths around the distributor cap and went through. It fired up straight away on the other side.'

L to R: Fulfilling a bit of U-boat fantasy.
Three Dutch blokes taking on the Kimberley to test their theory of 'Nobody in Australia needs a 4WD, just buy an old Commodore and drive'.

'So you're getting the towing journey of the Kimberley?' I chided.

'No, it's just the rivers. We've done all right. We've had two flat tyres and we lost the exhaust, but we put it back on. It's just a bit of an adventure and we are willing to take some risks.'

'So it's a bit of a bulletproof car?' I queried.

'Well I don't know, it takes on some wallabies and kangaroos, though.' He laughed through his clipped English. 'We are off to Fitzroy Crossing to get the car checked over. Then we are going down to Ningaloo reef to swim with the whale sharks and then find a job, big time.'

I wanted to know if jobs were needed to fund the major repairs.

'Nah, this baby is not going to give us any trouble. Nobody in Australia needs a 4WD, just buy an old Commodore and drive', he cackled.

'And maybe pack a snorkel for the river crossings?' I suggested.

'Well there is that, and a lot of beer.'

He did concede that it hadn't been all smooth sailing—tyres seemed to be the main issue. In all, they'd had 7 flats since leaving Melbourne. One of the other Dutch backpackers was a little more forthcoming when away from his mates. He quietly admitted that there had been a couple of times where he wasn't sure they were going to make it all the way through.

When I met the Dutch Commodore Team the worst of the Gibb was well and truly behind them; the last speed hump they had to get over was to cover the 140 k's to Fitzroy Crossing. Fortunately they had found a way to make this a little more challenging than it needed to be.

L to R: Of my two separate journeys around Australia each took a different route to Broome and yet both bikes clocked up 10 000 ks at roughly the same point. Freaky coincidence or just a freak that notices?

Northern Australia during the dry season: everything seems to be either burning or burnt. These are some of the more creative efforts urging people not to add to the problem.

'We are down to 10 litres of petrol, so the last part is going to be a bit exciting. This car is like us, it drinks a lot.' With that the Dutchman laughed and climbed back into the driver's seat.

The previous year's encounter with Dutch boys played in my mind while I rode further into the Kimberley. The road from Broome to the start of the Gibb was the last bit of tar I would see for a while. The tar provided a chance to relax before I got on the track. I'd heard way too many horror stories about the Gibb, so I tried to quell my concerns by reflecting on the fact that I had met 3 foreign city boys who had conquered the Gibb in a flaming station wagon! I focused on convincing myself that it would be a snap, I had the perfect bike for it, petrol was not going to be an issue and the rivers were reported to be running low, so I would be laughing—I hoped.

The only thing standing between my goal and me were a few bushfires. It seems that north of Broome during the dry season everything is either burning or burnt. While riding past yet another scrub fire I glanced down to check my gauges, and through the smoke the odometer was reading 10 000 k's. In between thoughts of 'oh my aching butt cheeks', it dawned on me that I had hit the 10 000 k mark when I got to Broome last year, a freaky coincidence considering that on that journey I got to Broome via Highway One. I put the nerdy train-spotting stuff to one side and got back to the business of choking and gagging on the smoke.

A few hours and countless fires later, I was standing at the end of the bitumen, the point at which the Gibb started in earnest. I was convinced it was going to be the toughest ride of the journey, and to put me at ease there was the obligatory safety sign saying:

WARNING
LOOSE SURFACE
DUST
CORRUGATIONS.

Strangely there was no advice about bloody great river crossings, bushfires, crocs or any of the other stuff that can really put a dent in your day.

The start of the Gibb symbolised the start of the Kimberley proper for me. (The map says it is back near Broome, but for my money it really kicks in at the Gibb.) The Kimberley is a whopping 420 000-square kilometre region, an area that equates to about 3 times the size of England. A big chunk of turf, with only about 25 000 people dotted across it. And if you believe the marketing spiel, this makes it one of the least populated places on the planet. The Gibb is the only road through the heart of the Kimberley, making it a vital link for the people spread throughout. The Gibb, though, is also becoming a tourist route. Until recently, 'remoteness' had saved the Kimberley from mass tourism, but marketers have given it the title of the 'Last Frontier', which means that a flood of rubberneckers will probably soon result in it losing all the characteristics that earned it that title.

The beauty of the region starts abruptly at a reef. The Devonian Reef was a hangover from when the place was under water 350 million years ago. The reef, or if you want to get technical about it, limestone cliffs, burst up out of the flat landscape, a 60-metre grey-blue wall threatening to block the road and conceal what

L to R: Queen Victoria's Head, Gibb River Road.
A bit of spooky Boab action…they look less menacing with leaves.

lay on the other side. Just as the Gibb looked like it was about to turn into a 'dead-end', the road veered left, snaked through a gap and passed under an outcrop called Queen Victoria's Head, which from a distance did look a little like the old fun-bag.

Once through the reef, the track starts to wind its way up through the King Leopold Ranges, a hummocky landscape peppered with termite mounds and boab trees. The boabs were comforting as I have a few guide posts when I am travelling, including:
- I know I'm in the Kimberley because there are wall-to-wall boab trees.
- I know I'm north of the Tropic because there are termite mounds everywhere.
- I know I'm north of Broome because everything is on fire.

(With navigation points like these, it's not surprising that few people volunteer to travel through WA with me. Generally it's only those who have no qualms about finding their way back home with the aid of State Rescue stick their hands up.)

Despite the 'Last Frontier' marketing push, the Kimberley is yet to feel the full force of mass tourism. Dirt roads and the wet season have so far discouraged the onslaught of resorts, hotels and motels. There is of course a downside to all this— it's a bit of a bugger to find a bed at night. Much of the accommodation is in the form of makeshift safari camps, erected for each dry season to take advantage of the influx of tourism. Such a camp was to be my home for the night, but as usual I was running late and darkness had beaten me before I got to my destination.

Grey nomads in the making: Charmaine and Russ. To see what the life of a grey nomad was like they decided to find out by running a Safari camp.

Looking for a campsite in the dark is always a bit of a challenge, consequently I spent the early evening riding into various camping grounds, interrupting people's meals and asking them if I had found Imintji Safari Camp. Eventually I got a 'yes'.

It turned out I would have found the camp a lot easier if I had used my ears instead of my eyes: a king brown snake had slithered through the camp earlier on and caused a decent commotion among the campers. Sadly I had missed the action, but I did get to enjoy follow-up discussions about how 'snake-proof' everyone considered their tents to be. Fortunately for me the snake was last seen vanishing down a hole at the other end of the camp. I happily calculated it had about 20 tents to work through before it got to mine. My smugness was short-lived. Charmaine, the camp host, regretted that she had to inform me that the zip on my tent was broken across the bottom—the resultant gap was big enough to parallel park a dozen king browns in.

Besides myself—and the snake—the remainder of the guests were from a 4WD bus tour. I was a bit of a novelty act for this group of retirees who had been wrestling with the rigours of the outback in air-conditioned, cloth-seated, surround-sound luxury. After responding to 'You're not doing this track on a bike?' 20 times over, I got on with the business of unloading the bike and chasing up some food.

Charmaine and husband Russ, besides being camp hosts, were also 'grey nomads' in training. They had recently joined the ranks of the retired and decided they would build themselves a camper bus to spend a few years travelling around Australia in. The bus building was put on hold when they spotted a newspaper advertisement that read:

'WANTED: ADVENTUROUS COUPLES TO WORK IN THE KIMBERLEY'.

They thought, 'Why not get a bit of a taste of the life we will soon be living', and applied. Charmaine and Russ got a bigger mouthful than they had bargained on; it was a bit like signing up for the Scouts then suddenly finding themselves in the SAS. The camp hosting was hard work, perhaps largely because everything was done out of tents and humpies. Every day they had to clean out the tents and change the linen on the beds inside them—there was no luxury of sending the sheets out to a laundry service, it was all done 'in-house' on a couple of washing machines parked in a lean-to under a tree. When the 'rooms' had been dealt with then came the issue of feeding the guests. The kitchen-cum-dinning area was also housed in a tent, admittedly a little bigger and flasher, featuring a couple of battery powered lights hanging from the centre poles and a fridge parked in a corner. The showers and toilets were also bits of bush craftsmanship that required continual maintenance.

> They were seen as the 'grandparents' on the camp circuit as all the other hosts were under 30.

Russ and Charmaine had bitten off a lot of cooking, cleaning and maintenance in a very hot and demanding environment, which was particularly impressive considering they were as old as the retirees they were looking after. They were seen as the 'grandparents' on the camp circuit as all the other hosts were under 30. Considering the workload, and their senior status, I wanted to know if they were finding it a hard gig.

'It depends what you call hard. No, not really. It's long hours but it's very rewarding with all the people you meet. People from different cultures, different countries, it's just great', a smiling Charmaine told me.

With so much work involved I was surprised they had much time left to meet and greet.

'We have a little bit of time in the mornings, but not when a tour comes in at night—we are just flat stick getting them into their tents and getting tea ready.'

When they did get a chance to sit and chat with the punters, it often left them with food for thought.

'We had a young couple come through early in the season, the guy would have been about 19, and it was the first time they had ever been away from mum and dad. The first thing he said to Russell was:

'Where's your cabin?'

'Don't have one.'

'Where do you sleep?'

'In a tent.'

'Oh you don't, you're not sleeping in a tent for 6 months are you?'

'Yeah.'

'Where's the TV?'

'Don't have one.'

'Where's the radio?'

'Don't have one of those either.'

'Oh, blow that!'

The young bloke couldn't believe it, but we don't miss it, not at all. Who wants to listen to violence and crime and all that going on out in the world, when we have got all this beautiful peace and quiet and nature', Charmaine said.

I looked down at my microphone and said, 'You are making me think I should go down to the creek and chuck my recorder'.

She laughed and assured me I wasn't part of the problem. I stood up, bade them goodnight and made for my tent, wondering if there might be scaly company waiting for me.

24

Dirt surfing

I awoke 'snake-free'. It was a good start. After doing the usual morning things to get ready, I asked Charmaine one last question before I departed.

'How would I sum up the Kimberley?' repeated Charmaine, pausing to gather her thoughts.

'It's magic, every turn in the road is a new scene. We've hardly taken any photos because we reckon they just won't do it justice.'

You don't get recommendations much better than that, so I smiled at her, waved goodbye and rode off. The track twisted out across grassland and cut through a deep creek crossing before it joined back up with the Gibb. I had a newfound respect for the Gibb thanks to a story Russell had told me. He said he had met a motorcycling husband and wife team a few weeks earlier. They arrived

at the camp exhausted, with the woman vowing she could not ride any further after having come off the bike 10 times in the one day. The husband went into support mode and spent the evening at the safari camp trying to talk his partner's confidence up. Russell said the psych session worked because she got back on the bike the next morning and continued along the Gibb.

Such a high number of falls sounded a little extreme, and I hoped they were because her riding skills were just not up to the task, rather than the road ahead being an outback nightmare. As I headed further east her difficulties started to gain some credence—the road was covered in thick loamy sand, deeper in the middle where passing traffic had banked it into fat ridges. On hills and in corners the soil was rolled into vicious corrugations, a little bigger on every new incline and turn. Even though the Kimberley was upping the ante I was enjoying the ride.

My confidence in the bike, and my riding, had grown dramatically since the central Australian tyre debacle. Tyres, though, were again on my mind—the rear one was balding quickly. The rear wheel spinning in the loam was quickly chewing away the tread, which was disappearing at a rate that made me wonder if there was enough there to get me to the end of the Gibb.

While I was busily calculating tyre life versus k's of the Gibb remaining, I shot past a sign saying 'Galvans Gorge'. I had been given a tip to have a gander at this gorge and the thing I liked most about the description of the place was that it didn't require a marathon hike to get

Although that might sound like I am a lazy bugger, the issue is that motorcycle suits are not the most comfortable things to trek in, as they weigh a tonne and quickly begin to develop sauna-like qualities.

199

into it. Although that might sound like I am a lazy bugger, the issue is that motorcycle suits are not the most comfortable things to trek in, as they weigh a tonne and quickly begin to develop sauna-like qualities. I could make life easier by leaving the jacket with the bike but, knowing my luck, it would get swiped by the one dishonest bastard in a 2000-k radius. Security was a continual issue with the bike; whenever I parked somewhere it was always in the back of my mind that a crowbar could easily liberate my equipment boxes, a prospect that would finish my journey. Then again, you can worry about everything and still get done over, so I parked the bike and headed for the gorge.

Soon I was wading through thigh-high grass, very happy about being in armoured riding pants and boots. The protection even made me feel a little bolshie about the thought of crossing paths with any snake I might meet. As I was thinking 'bring it on', I turned a corner and came across a metre long Monitor lizard sunning itself across the path—my desire for a stoush evaporated immediately. The lizard was unperturbed by my arrival and continued sunning itself. Either it was hoping I hadn't seen it or it was just one badass lizard and none too worried about my presence. It didn't look like it needed to worry; its scales shone like chain mail and its claws could make a can opener redundant. I remembered being told that when lizards are scared they seek refuge at the top of the nearest object and they are not fussy about whether that's a tree or a human. The thought of being used as a fleshy scratching post encouraged me to crouch low and consider my options. While contemplating my next move I dragged out the camera—the terrifying sound of the camera clicking solved my impasse by sending lizard scurrying on its way.

Beyond the lizard, the trail followed the banks of a creek for a short distance before petering out into a large pool. I am not sure why the title of 'gorge' was bestowed upon the spot—in my books if was more of a waterhole. At one end it was hemmed in by a cliff with a large boab precariously positioned at the top. Boabs shed their leaves for the dry, and the gnarly naked arms reaching out from fat squat trunks always make them look a little menacing. This one was no different and overlooked the waterhole with malign intent.

The pond was fed by a small waterfall running down the cliff. Despite the baleful boab, it was a peaceful spot to soak up a bit of nature and it didn't take long for the cool inviting water to get the better of me—I shed my kit and hit the water. I told myself, 'if I get sprung skinny dipping, I get sprung', which in hindsight was highly probable, considering it was a tour driver who recommended I should check out the gorge. The water was cooler than I anticipated and my hot-blooded enthusiasm was quickly tempered. With a clearer mind I decided that being sprung solo skinny-dipping at a tourist spot was going to look a little sad, and also a little porno, so I made for my clothes. As I left the gorge I bumped into 3 young German travellers—had they arrived 5 minutes earlier I fear they may have been put off nature for life. Still, it was probably better to upset young backpackers than cause a series of coronaries among the senior citizens who visited the gorge.

After an hour of gorge action, I was back on the bike and heading for Mount Barnett, my one fuel and lolly stop on the Gibb.

The Gibb can be a battle of attrition, and even if you are winning the chances are that some bastard will come along and tell you it isn't going to last for long. My bout of negative reinforcement came at a river crossing the following day. I needed some pseudo action shots for a TV yarn, so I decided that a bit of footage of me barrelling through a water crossing would be the go. Videoing, at any time, is a labour-intensive process, and trying to video yourself doing something doubles the pain factor and requires a lot of tripod moving to get umpteen different angles. Recording yourself is a wacky task and it usually attracts attention, so naturally the only vehicle I had seen all morning turned up just in time to watch the spectacle. After a few crossings, and tripod moves, the spectator came over for a chat.

After we got through the details of why I was so mindlessly obsessed with videoing myself, the conversation turned to the Gibb. He wanted to know which way I was heading, and when I told him 'Kununurra' he eyed the bike over and said:

'Where's your spare tyre?'

'Don't have one mate', I responded.

'You're stuffed. You'll never get past the Kalumburu turn-off. I had 3 blow-outs between there and the Durack', he told me with surprising glee.

For some reason this bloke was intent on giving me a hard time, and after a few more minutes I was starting to suspect he was your garden-variety dickhead. I tried to tell him I had puncture repair equipment and that riding a bike was a bit different to driving a car.

'You can dodge a lot of the things that give you blow-outs', I said. I politely substantiated the point by saying I was on my second lap of the country and was yet to get a flat.

As I finished my 'coming straight back at you' spiel, another traveller pulled up and lent out his window and joined our roadside conversation.

Mr Positive was obviously not happy with my retort and was intent on making a point by asking the newcomer what the track was like beyond the Kalumburu turn-off.

'Oh it was hell! I came through there yesterday and I trashed 2 tyres', the newcomer replied.

I couldn't believe it, a total stranger, within a minute of turning up, had blurted out an answer that sounded like it'd been scripted by Positive Boy (who looked visibly pleased with this new supporting information).

Keen not to be infected by these blokes' inability to dodge tyre-wrecking obstacles, I said, 'Thanks for the tips, but if I hang around chatting for any longer I won't get to wreck some rubber myself', then grabbed my helmet and left.

The bloke had got under my skin. I had enough self-doubt without some other nazi adding to it. The negativity, though, had got to me, and as I rode towards the Kalumburu turn-off I became increasingly sensitive about any changes in the track. Every time it got a bit rougher I wondered if this was the start of the tough section he had gloated over.

I made it to the turn-off without anything going pear-shaped, and recalling that Mr Positive did say 'after the turn-off', I decided a breather might be a good plan before beginning the onslaught. Also parked at the intersection was an all-wheel drive tour I had met a few days earlier. They were having lunch and invited me to join them. As I was in my usual lunch-less state, it was an offer my stomach insisted on accepting.

Leading the tour were guides Rick and Wren. Wren had been doing tours through the Kimberley for a couple of years; she was in her 20s and had been driving trucks before turning her hand at outback tours. Rick had moved up recently from down south and this was his first season in the Kimberley, but he had been driving tours for 10 years in various places in WA. Tour guiding looked like a bit of a blast and I got chatting with Rick about whether it was the fun I imagined.

'It is', he said. 'It's wonderful hearing people from other nations saying "Wow". We take it for granted. A lot of Australians say, "We'll get there one day" and they never do. But when people do it this way, on the tour, they say, "Great, fantastic, why don't more people do it?" This time around I'm lucky, I've got mostly Australians and they are doing something they have always wanted to do and I am lucky enough to take them, so it's great.'

Wren and Rick, Kimberley tour guides, arguably one of the better jobs on the planet.

'Do you think the outback is an eye-opener for a lot of Australians?' I asked.

'It is', Rick replied. 'If you go back 10 years ago it was completely different—you didn't have the facilities like the camping grounds, the safari camps. It used to be much harder. The younger travellers still want to do it the hard way, but when you get a little older a few home comforts are good.'

I thought to myself that the camps were also good for the odd clown on a motorbike who decided to carry recording equipment instead of survival stuff like food. Besides helping mugs like me out of a squeeze, I wondered if the camps and facilities were making the bush smaller and more accessible.

'Yes and no. I don't think you're ever going to make the bush smaller, but more accessible, yeah. The guys I have on board want to do it, but perhaps they don't want to have the pressure of owning their own 4WD, maintaining it, or worrying about where they are going or the time frames involved in getting there. So they give us the responsibility to do it for them, and that's great.'

Rick seemed genuinely enthused about what he was doing for a living, which was remarkable considering he had been doing it for 10 years—a long time in the leisure industry, as it often comes at a high personal expense.

'It's one of those lifestyles where you've got to put everything to ground, there is no such thing as a weekend off. It teaches you a lot about yourself as well as about human nature, because in dealing with the travellers you're actually dealing with yourself, and over the last 10 years I've grown a lot. I've still got a fair bit to do, but I'm still learning', Rick said candidly.

It was a surprisingly frank comment from a bloke, especially given that Rick looked about 40 and didn't fit the 'young and idealistic' tag. Maybe it's both the inner and outer journey that makes travellers so obsessed about keeping on the road. I've been guilty of the same myself, and occasionally I've wondered if the best place for me to work would be in an adventure travel business. The thing I feared, though, was that I might get bored with it quickly and then find it hard to maintain enthusiasm. I wondered if it was the same for Rick.

'Yes and no. It's the response from the group that's the winner. This group, they've been around, a lot of them have travelled around the world, but they say, "Wow, this is fantastic, this is why I've come". That feedback makes you more enthusiastic.'

'Are you surprised about Australians being ignorant of what lies beyond the urban fringe?' I asked.

'Oh yeah, but this is the part of the deal with what I do. I show them and give them all this knowledge that I've picked up and they start to say, "Hey, maybe when we get back we can buy or hire a 4WD or a camper van, and go out and do our own thing". Anybody can do it, all they've got to do is take the right precautions and the right vehicles and they can do it', said Rick.

As I looked over Rick's shoulder to the track beyond the Kalumburu turn-off, I was not entirely convinced about his statement 'anyone can do it'. The track was an angry seething mess, chopped up with a swell of corrugations that broke on the shore of the turn-off. Rick and his driving partner Wren both assured me the next section wasn't as bad as everyone said it was.

'It's a bit rough, but you'll be right, you're looking at the worst bit', Wren told me.

I thanked them for lunch and took the opportunity to top up my waterbag from the truck's water tank before departing. The tour group stood around to wave me off; the looks in their eyes seemed to express everything from admiration to incredulity that I could be so daft as to be out there on a bike. I punched the start button and gave the engine a good rev—a bit of yob behaviour to encourage the undecided to put me in the 'daft' category. I gave a nod goodbye, dropped the clutch and quickly started stabbing my way up through the gears. I wanted to be up to ramming speed by the time I hit the corrugations so I could skim over the top of them. 'Skim' may have been the theory but it certainly wasn't the motion. I felt like I was a rag doll strapped to the back of a jackhammer. The corrugations were vicious, designed in hell and installed in the Kimberley. On a bike you can usually find a smooth patch of road and escape the pounding, but there was to be no hiding on this day. The rolling dirt surf stretched from the boab bases on one side to the termite mounds on the other. The chop thrashed the bike around like an abusive parent, rattling my brain in the process. The handlebars were hammering into my palms; I was convinced my arms would be a smaller sleeve size by the end of the day.

The further I rode the wider the track became, pushed out at the edges by vehicles desperate to hug the shoulders of the road, each driver vainly hoping that it might be smoother at the extremities. In my own hunt for a calmer passage I was using every inch of the track, regularly darting from one shoulder across to the other, chasing smooth, flat mirages that always dissipated when I got there. Anyone following me would have kept their distance, fearing my erratic manoeuvres were fuelled by either alcohol or lunacy. I wanted desperately to stop and rest, catch my breath, but I feared I would never be able to get back up to speed again (at least not without going through a lot of pain). Stopping for a drink would also have been welcome, as I was sweating buckets and desperate for some water. Normally I drink while I ride but there was no way I could take a hand off the handlebars to grab my drinking tube and shove it in my mouth, so water would have to wait. I continued to zig and zag, dodge and weave, ever conscious of avoiding the evil looking rocks that littered and jutted up out of the road. I was beginning to understand how the blokes I had met earlier had got so many blow-outs on this section but thoughts of Positive Boy raised my blood pressure and made me more determined to keep my 'no flat' status.

After what seemed like an eternity in one way, but a few fleeting moments in another, the corrugations ended as abruptly as they started. The road stopped climbing and was now winding over a harder surface that was less predisposed to being chopped up. I stopped for a drink and to check that nothing had been damaged or broken free, certain that something would have gone astray after such extended and extreme abuse. Remarkably, everything was in order. I gulped down a bit more water and set about getting to my next obstacle—the Durack River.

The Durack has a reputation as being one of the main obstacles along the Gibb. Fortunately the previous wet had not been that big and the reports were that there was little water flowing through it. I pulled up on the river bank and gazed at the crossing. It seemed that for once the reports were right and the crossing should be a snap. Arriving not long after me was a Land Cruiser, which I had caught glimpses of for the last half-hour as it followed in my cloud of billowing dust. I was pleased they had finally caught up—I wanted to use them as a guinea pig and watch them do the river crossing. That way I could see if there were any nasty holes to avoid. The cruiser stopped next to me and the driver leaned out the window and said.

'You weren't dragging your heels, mate.'

Unsure if he was having a go, I replied, 'Neither were you'. He chuckled and stared at me with a grin. I told him it was all smoke and mirrors, a quick dance that weaved its way along the smoothest path I could find. However, I was going to be limited to a crawl through the crossing, once I had seen the route he was taking.

He gave a wry smile, said goodbye and bounced off through the crossing. The riverbed looked rocky, but fortunately the water was not too deep. I love water crossings, the bike was built for them and so was my riding suit—everything was designed to deal with a decent dunking, everything except the flaming recording equipment and computer. The equipment boxes were supposed to be watertight, but I figured coming off in a river and turning them into mini

U-boats was probably not what the manufacturers had in mind when they described them as 'watertight'. Most water crossings were usually short and not a problem. With these I took the approach that if it went pear-shaped halfway through I could twist the throttle violently, career out of the water and crash safely on dry land. However, wider crossings required a little more strategy and diligence, so I tried to remember the path the Cruiser took. I started the bike, aimed the front wheel, and launched into the river. The bike began pogo-ing across the waterway, large rocks slapped the wheels from side to side and kicked the bike up and down. The rocks felt large, the sort of small round boulders any sane rider would avoid on dry land, however the camouflage of water took away the choice of avoidance. All I could do was hit them, hold on, and let the bike surge through the water.

Before I had a chance to settle into the leap-frogging rhythm, the bike had leapt out on the other side of the crossing and we were charging up the river bank. The crossing had been a breeze. I breathed a sigh of relief and ticked off another Gibb hurdle; the only one that remained was the Pentacost River. Unfortunately I knew first-hand how big an obstacle that was going to be—I had ridden up to it from the eastern side the year before. It was a proper crossing, deep in patches and a couple of hundred metres wide.

The following day I was unloading my gear on the banks of the Gibb's final challenge. Once I crossed the river I would be on the home straight, with the toughest part of the entire journey conquered. Because the crossing was so wide I was not going to risk carrying my equipment: it would be just begging for trouble. I decided I would try and bum a lift for my gear in a passing 4WD and I would follow on the bike.

In the end I was saved the hassle of hitching when the people I stayed with the night before offered to ferry my gear across the river. Anne and her 6-year-old son Brodie were to follow me down to the river in their tray-top, stopping on the way to pick up the mail. As the Gibb would be hell on a postie bike, the postal service decided the best way to go about it was by air, and the day of my crossing was also 'mail plane day'. We arrived at the airstrip at 9.30 am and the small plane landed moments later. It taxied over to us and Anne greeted the pilot. After a brief exchange of parcels the aircraft was on its way again. The postie offered a two-way mail service and also a courier service—for a price they would bring out anything

else that could be fitted into the plane. (The size of the plane is crucial. One of the stations in the Kimberley, Kachana, is only accessible by plane. The Swiss bloke who owns it has spent years building it from items flown out by plane—the story was that if it didn't fit in the plane, then it wasn't used in the building.)

We left the airstrip and began snaking our way down towards the river. The Pentacost was in the bottom of a wide valley and easy to see from a long way off. In the wet it was even easier to see. Anne's husband told me tales of how the river would swell to fantastic proportions, rising by 15 metres and spreading out kilometres wide. He said that now—the middle of the dry season—was the time to cross it and I would be fine. Just when I was starting to feel confident about getting across he added, 'There's also a croc that lives a little bit down from the crossing. It's a reasonable size, about 5 metres long. But it doesn't much like the noise of the cars going through the river, so it keeps its distance.' I was pleased to hear the bloody thing had some road sense, nonetheless I was determined to make as much noise as possible and stay on the bike.

I arrived at the edge of the river and waited for Anne to catch up. I had enough waiting time for my mind to start building the crossing into Red Sea proportions. Unfortunately Moses' number wasn't in the sat. phone, so I resigned myself to learning what I could from a pair of 4WDs that were about to cross it. My hope of being educated was shattered when one of the blokes ambled over and asked me what I knew about the ford. I told him I'd heard it was reasonably shallow, but to be wary of a deeper hole near the other side. In the course of our

L to R: The mail plane, the only sane way to deliver the post throughout the Kimberley. Arguably the biggest obstacle along the Gibb River Road—the Pentacost River.

conversation his travelling buddy also wandered over and the discussion turned to which one of them was going to tackle it first. It began boiling down to who had the best rapport with their insurance broker.

I left them to it and started unloading luggage in readiness for Anne's arrival. By the time my gear was unclipped, team 4WD had made a decision and the first vehicle launched into the river. They easily made it to the halfway point where the water was only lapping about the axles. The further they went the more my confidence grew; not only was the river not that deep but I figured the submerged track was relatively smooth, as the 4WD wasn't violently pitching about. However, as the Cruiser neared the other side its bonnet plunged violently downwards, taking my hopes with it. The wheels vanished under the water and the truck began heaving from side to side like a cork bobbing down a river, as opposed to a 4WD conquering one. As they continued to bounce and surge through the water it began to dawn on me that days and days of battling the road had been for nothing—the last 30 metres of the crossing were impassable for me, and I was going to be forced to do the biggest U-turn of my life.

Before depression had a chance to truly set in, the Cruiser began climbing out of the river and was soon dripping dry on the opposite bank. Back on my side of the river it seemed I was not alone in my concerns, the other half of the 4WD expedition were also looking a little ashen. Discussions were held and the first car was hailed on the CB. After information was traded about the relative state of underwear, it was concluded that it wasn't that bad and that they would have been better off if they'd stuck to the up-river side of the track. The second car drove into the crossing, following a line slightly more upstream from his mate. Nearing the other side the driver slowed, veered a little more to the right and headed into the final section, the car dropped, the left side more than the right, but not as deep as the previous submariners. The second driver had found the sweet spot and bobbed his way through and back on to dry land. This swung my mood again, and I now felt more up-beat and fancied my chances of getting through.

Anne arrived and when I told her what I saw she said, 'Yeah, I was going to say, you might be better off on the high side'. The 'high side' was the key, and now I was confident that I could conquer it, I decided to complicate things further: I was going to do the crossing several times. Most sane bike riders would be happy just to get through it once, but I had the greedy beast called TV to satisfy and the crossing would provide some good footage to shove in a story. Getting the vision, though, would require a couple of different camera angles and therefore a couple of crossings.

I assembled the camera and tripod and gave Anne 5 minutes more training on the equipment than I'd ever received. I grabbed my helmet and set off. The first half of the crossing was a snap, so I went reasonably quickly to ensure the best possible splash factor for the camera. I slowed as I neared the hole on the other side and steered a little upstream, preparing to make like the Nautilus. Before I was ready the front wheel vanished and the rest of the bike plummeted quickly after it. The front-end departure was brief as it reared back up after slamming into a large submerged rock and the wheel porpoised briefly above the water before diving back under to continue its voyage. I ricocheted from rock to rock, cursing and wondering if BMW had ever considered fitting sonar to their bikes. The size of the rocks I was clambering over determined how much of the bike was submerged. Most of the time the cylinders were below the waterline and I winced at the thought of my red-hot engine trawling through the cold water. I suspected that was treatment probably not recommended by the Bavarians. Despite the torment I was putting the engine through, the bike continued to bounce along underneath until it flicked over one last rock and, like a landed fish, flung out of the water and onto the bank.

I rode up the track a little, just long enough to wipe the smile off my face, before turning around to do the return trip. I aimed at the drip marks showing where I had just come out and plunged back into the water. This time I knew what I was in for and let the bike move freely around underneath me. The second time around the deep section seemed less of an obstacle. The bike bounced up and down and from side to side and by the time I was into the pogo rhythm we were out of the hole and  surging across the shallows of the river. A small rocky section in the middle forced me to slow a little before putting on more speed and aiming towards Anne. I burst out of the river and shot past Anne, the final hurdle of the Gibb was conquered.

I rode back around to Anne, who greeted me nonchalantly, 'See, I told you it would be easy.'

I smiled and climbed off the bike, keen to check the vision. Anne had done a great job, but there was one minor problem. Brodie had got into some of the shots—it just didn't look like the conquest of a croc-infested river with a 6-year-old wandering around at the water's edge. Television and vanity got the better of me and I made one of the worst decisions of the entire journey.

'I'll do it again', I announced stoically, grabbing my helmet and bolting. This time I rode a little faster, racheting up further my 'the more water for the camera the better' dictum. The bravado wore off at the deep section. I slowed, picked my line and let the bike begin its wallowing and bouncing. A short time later we were again up on the bank, briefly drip-drying before I turned and headed back into the water for the fourth time. By this time my boots were reasonably full, which must have caused my brain to become waterlogged—that is the only way I can explain the moronic decision that followed. I decided I should try and enter the river a little more downstream, as it might make for some more dramatic pictures. I was right on that count, I just didn't count on how dramatic.

I entered the river expecting the front wheel to disappear under the water, but the thing I wasn't banking on was the petrol tank following after it. Water was lapping high, too high. I knew instantly that I was in the bad books as far as the motor was concerned—it was going to be sucking in water at this depth. The only option was to get out quickly. I steered towards the shallower side, but getting there was a problem. The rocks in this section were a lot bigger and the bike launched from one to the next like a demented frog. The third leap bucked me, I slipped off the footpegs and the show came to a grinding halt. Stopping was a bad plan, very bad. Water quickly started banking up on my left-hand side, the air intake side, and the engine was starting to falter, missing, gurgling, spluttering. I scrambled to get back on the pegs. If I didn't get the bike out of the hole before the engine died I would be thoroughly stuffed—there was no way I would be able to push it out over the rocks, or push it against the current. I leapt back into position, dropped the clutch and fed it a fistful. It burped, farted and coughed before summoning up the strength to drag us both forward. We rolled over a couple more large rocks before it limped out off the hole and died in the shallows.

I didn't need a mechanic's report to know what had happened: buckets of water had gone down the air box and into the engine. I jumped off the bike and set about removing the drain plug from the air box. Naturally I couldn't get it straight out, the bike wanted blood for the torture I had put it through and a burnt offering was made in the form of my hand slipping off the plug and searing on the exhaust pipe beneath it. Satisfied with my fleshy token, the plug yielded and water gushed out of the air box and back into the river where it belonged.

I had cleared and drained the airways but the next part was the tough part—starting the heart. After copping a lung full, the bike was understandably reticent to start. For what seemed like an eternity of pleading with the bike, turning the engine over and cursing my stupidity, the engine eventually spluttered back into life. There was a pulse, but the way it was coughing and hacking I suspected it should have been on life support. After a minute of gently massaging the throttle, the engine fell slowly back into a more normal rhythm, enough for me to try riding through the shallows back to Anne and Brodie. As each metre passed the health of the bike improved; it seemed roused by the approach of dry land.

Soon we were parked on the bank and I was inspecting the damage, checking where the water had got in. There is an old adage that oil and water don't mix, and as I crouched next to the bike, peering intently into the oil inspection window, the maxim was proven true. The oil had turned milky white—my stupidity had resulted in water getting in with the engine oil. I was in deep shit.

Anne saw the blood drain from my face and asked what was wrong. I explained that if I tried riding with water mixed in with the oil there was a good chance the engine would unload its undies in a big way. Anne kindly volunteered to return home and see if they had some replacement oil. The hour-long wait for her return gave me plenty of time to beat up on myself over my stupidity—not only did I repeatedly put the bike at risk but there were things I could have done to minimise the outcome. The first would have been stick to the same route each time and the second involved fitting a snorkel to the air intake. I had carried a pipe for 1000 k's, expressly for this purpose, and when it got to the biggest crossing I did not bother to put it in. What was I thinking? Had millions of corrugations finally rattled a few synapses loose? Had months on the road made me blasé and reckless? Was I now numb to how remote I was? Did I no longer appreciate just how isolated I'd be if it all went pear-shaped?

Anne eventually returned, unfortunately with bad news—they had no engine oil. I had no other choice than to attempt to ride to Kununurra and change it there. All that lay between me and Kununurra was one more crossing of the Pentacost, 120 k's and a lot of prayers.

ROBYN AND LISA

'We started out in Katherine Gorge and we've ridden to Kununurra, and now we are riding down the Gibb River Road for Broome, for a swim basically', said Robyn, astride her mountain bike.

I looked at the two women a little incredulously and asked, 'So how long are you giving yourselves to ride from Katherine to Broome?'

Lisa and Robyn.

'All together it's working out to be 24 days', Lisa replied.

Considering they were on pushies, it sounded like a very short time and I asked if the journey was still on schedule.

'Oh yeah, right on schedule', Lisa said emphatically.

It was pretty good going, especially considering their journey was 1500 k's, 700 of which was on the rutted-out mess that is the Gibb River Road. Robyn, though, was no stranger to long hauls on pushbikes; she had once ridden from Adelaide to Perth. I suggested to her that this was the wrong way around, as she would have been riding into a headwind all the time. She said she realised that after she started but still stuck it out.

I surveyed their bikes, which were weighed down with canvas panniers at each end, but surprisingly they weren't as bulky as I expected. I commented that it must be hard travelling on a bike, as they had to carry everything to ensure they could be self-sufficient.

'Definitely, and it also does wondrous things to your bottom', laughed Robyn.

Lisa was quick to chip in with some other specifics, 'But it also makes it so simple'. I raised my eyebrows questioningly before she continued, 'Honestly, all you've got to carry is a tent, some clothes, some food and water. You don't have to worry about anything else—petrol, tyres, generators.'

It was Robyn's turn to pick up the baton: 'We think it's hysterical when we get to a sign that says, "Warning, no fuel for 580 kilometres" and we think, "So?"'

'How long can you be self-suficient for?' I asked.

'Basically we can eat what we've got for 10 days and we refresh our water every day, or every second day', Robyn said.

'Well, you are certainly in a good part of the world for water', I commented.

'That's right, that's why we didn't do the Tanami, because we were uncertain about the water and we thought we might have to carry as much as 50 litres. We're working on it though. Next year, definitely', Robyn said in a serious tone that left me in no doubt that she would indeed be tackling it in 12 months time. Even so, I imagined it would be a more difficult challenge because the Tanami is sandier.

'It is supposed to be a mixture of sand, gravel and loose rock', Robyn agreed. 'This is a harder surface, supposedly.'

Lisa joined in, saying, 'But you don't actually know until you do it, because everybody reports it differently'.

Never a truer word said, I thought. One person's road is another's goat track. I had learnt that so many times over.

Besides track reports, I was wondering what the main concern was in tackling such a journey on a pushbike.

'Out here, probably safety or water. Personal safety is a bit freaky, if you are doing it on your own', Robyn said. She then described some of the precautions she took, which included waiting until there were no cars around each evening so she could race off into the scrub and set up a secret bush camp. She was a gutsy woman, especially when you considered she was only about the size of a poorly nourished sparrow. Nonetheless, she was a hardy traveller and there was a very determined look in her eye. Some of the rides she had done spoke volumes about her resolve; as well as the peddle to Perth she had also tackled the Oodnadatta track, the Mereenie loop and a few others in between.

'So you are not afraid of hopping on your push bike', I quipped.

'It beats a Melbourne winter', laughed Robyn, and rode off down the track.

25

Deserts one month, tropics the next

Four days and many litres of oil later I left Kununurra. After I had turned the bike into a U-boat on the Pentacost, I had visions of my engine self-destructing and a piston flying out and taking a leg off in the process. These fears, fortunately, were unfounded; in fact, all traces of water had vanished from the engine oil by the time I limped into Kununurra from the Gibb. Nonetheless, I took no risks, booked into a garage and started pulling things apart. The engine had its oil dumped, then flushed out with a fresh lot, then dumped again and then replaced with more fresh stuff. I checked all the other fluids for traces of water and then put it all back

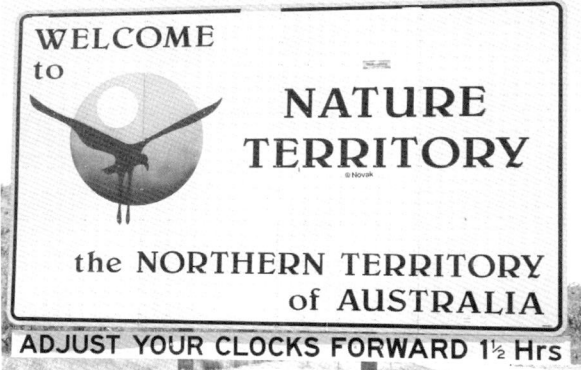

L to R: A bit of primordial swamp action on the edge of Kununurra.
Losing an hour and a half is never as much fun as gaining it…except in the Territory where no speed limits mean you can have a fat old time trying to make up for lost time.

together. Other than bike repairs, and a bit of 'giving thanks' for being let off with no apparent damage, the rest of the time in Kununurra was spent collating, editing and filing stories. After several days of sweating in front of the laptop I had caught up a bit, but I was still plagued with the problem that what I was filing was still way behind where I actually was. The only distraction I had from the computer was a desperate attempt to kill the stench that my boots had developed since being waterlogged. The smell was vicious and I tried to convince myself that it was from something within the boot and had nothing to do with my feet.

My final task before leaving town was to fill up. While I was drip feeding the tank up to the brim (I had learnt in the past that every drop counts) a bloke wandered up to me and said, 'So that's what a BMW 1150 GS looks like.' It seemed an odd way to start a chat, so I cautiously responded 'yes' and asked why the interest. He went on to tell me he owned a towing business and he had received a call to retrieve a damaged BM out on the Gibb. He said the bike was a long way back and the cost of the retrieval was going to be well over $1000. That was just the start of the bills for the bike owner, then he would have to cough up for it to be freighted to Darwin, the nearest repair shop, and then who knows how many thousands in repairs he would be up for once it got there. All in all, it was going to cost him a bomb. His story of misfortune was a good kick in the pants to remind me to be a little less cavalier.

I said goodbye to the recovery bloke and promptly went about forgetting my new vow to be more cautious. So quick was my promise-breaking that I did not even bother to make any inquiries about the road to my next destination, Top Springs. Nor did I bother to check if the roadhouse there was open! The ridiculous part was that I did not even need to go there—I was doing a 450 k detour just so some of my remaining journey to Darwin would be done on the dirt. The dirt detour meant I wouldn't be bored out of my scone on bitumen all the way, plus I had a better chance of finding interesting outback type stories if I got off the highway. Taking a moment to think about the extra 450 kilometres, I decided my brain was becoming warped by the two years of travelling. Distance no longer seemed to have much relevance, at least not when it came to travelling on the bitumen. Perhaps also my brain was in shutdown mode now that the psychological hurdles of the Great Central, the Gascoyne, the Pilbara and the Gibb had all been cleared. All I could focus on was getting to Darwin, which was to be the halfway point where I planned on taking a short break away from the bike and the computer. Normally I hate breaking up a journey, but my little melt-down in Broome proved to me that I could not keep going like this for a couple of years without something going astray. For once in my life I realised there are things more important than maintaining the flow of the journey.

So, with my brain in shutdown mode, and running late as usual, I belted out of Kununurra, confident of making up a bit of time when I got onto the unrestricted roads of the Northern Territory. Forty k's later I lost $1^{1/2}$ hours when I crossed the border. Although there was still the same amount of daylight left jumping my watch hands forward was encouragement to tuck behind the windscreen and embrace the lack of speed limits in the Territory.

Two hundred k's later I was turning off the tarmac and on to the Buchanan Highway. Highway is an optimistic description for a single lane dirt track. The route was carved out in the 1860s by a bloke, not surprisingly, called Buchanan who had a vision of linking the Kimberley and Queensland cattle runs. A hundred and forty years later the track was still in use, and I was questioning how much work had been done on it since Buchanan pegged it out. The surface was awash with loose gravel, evil round stuff that was like ball bearings to ride on. The slippery surface, and a bald rear tyre, made for tough going. If that wasn't enough, the only car I saw for the day decided my side of the road was far more preferable as it popped out of a blind corner and thundered towards me. After successfully playing chicken, Outback style, I wrestled the bike to a stop and when the dust settled I began reassessing why the hell I was on the Buchanan. It was lunacy; the bottom line was that I had more recordings than I had time to craft into stories and file. I just did not need to be out there.

  My misgivings were later muted when the scrub that hemmed in the track gave way to reveal the beauty of Stokes Range, a stunning rock-topped landscape with a river carving through the middle of it. Being the dry season, the river had retreated into a series of interconnected pools, all cool and still, tempting me to abandon the bike and escape the heat with a quick dip. Sights like the Stokes often left me feeling a little torn; in the couple of years on the road I had seen so many beautiful things but I never had anyone with me to share it with. The irony was that I did end up sharing it with people, but it was a one-way relationship through the website etc, with people I would never meet. At least I was sharing it, but for it to have resonance I knew that nothing beats having someone there with you at the time. So much of my travels overseas had been by myself, but out of all the journeys the most valuable ones were those I had undertaken with others. On my ride through Asia I was with my friends Justin and Simone—not only could we soak up the beauty of, say, the Himalayas together, but in 5 years time we could still talk about it. I've got some wicked photos from all my travelling, but the ones I value the most are those that I can reminisce over with someone else.

  The beauty of the Stokes soon passed and the range tapered out into a flat horizon. The road straightened and arrowed towards Victoria River Downs, a cattle station that was once the largest cattle property in the world. Cattle are still part of the deal, but to look at the homestead you could be forgiven for thinking that the place had been turned into an outback airport. Parked out the front were half a dozen helicopters and there were several more in various states of assembly in the surrounding sheds. Apparently the property was now also home to a company called Heli-muster—the name said it all. I wanted to stick my nose in and find out how they fared running such a business in the middle of the outback, but the sun was getting low and I needed to push on. Judging from the immaculate grounds and the number of aircraft, I suspected they fared pretty well.

L to R: The road to Top Springs, a 400 k dirt detour just because I couldn't cope with riding all the way to Darwin on the bitumen.

The front yard of Victoria River Downs, once the largest cattle property in the world…now arguably the Northern Territory's largest heliport.

The bustling mecca that is Top Springs.

I pulled into Top Springs just after sundown. In fact I shot straight past as I didn't see it—there was not a single light on. The place looked deserted and I began wondering why the hell I hadn't bothered to make some inquiries before riding out into the middle of nowhere. I berated myself. I had learnt nothing from the Gibb. Before I set about dragging the tent out and establishing a foodless camp on the side of the road, I decided to venture around the back of the roadhouse and do one last check for signs of life. As I rounded the back I spotted a light on inside the building; there was hope yet for my grumbling stomach and near-empty petrol tank. As well as the light, there were sounds of people chatting. I parked the bike and wandered over to the building. A tall portly bloke came out and greeted me with 'G'day', in an American accent.

'You are lucky you turned up today,' he twanged away. 'We've been shut for the last few days. The manager quit, upped stumps and took the staff with him.'

The American sounding bloke and a couple of other bods were an emergency crew, trucked in from the local Aboriginal community a few hundred k's away.

'We've just opened the place back up, so hopefully there is a bit of food about and we might even be able to sort you out with a bunk.' He smiled and walked off to see what he could do, leaving me to think that I had managed to avoid another disaster by the skin of my teeth. I needed to get to Darwin before my luck ran out.

For a week I picked my way slowly towards Darwin. En route I paused at the turn-off to the Carpentaria Highway where a signpost told me that Cape Crawford was only 270 k's away. Cape Crawford was pretty much at the start of the Gulf Track, which was the route I would be using later to head east. I snorted at the distance on the signpost—it bore no relevance to the distance I had to cover before I got there. Not only did I have to go to Darwin first but from there I also had to detour a bit more before I reached the Gulf. It would be fair to say, in fact, that it was quite a bit of a detour, as it involved returning to Adelaide and then working my way back north to the Gulf via Birdsville and Mount Isa. It was going to take thousand and thousands of k's to get to a place that was only 3 hours down the track from where I sat! The concept was totally nuts.

  I stared down along the track and reassured myself there was some method to my madness. I needed to return to Adelaide for multiple reasons. Firstly, for my own well-being I needed to get off the road and rest and recharge—the Broome episode continued to weigh on me. I knew I needed some time out. This seemed like an odd thing to think; after all I had the best job in the world. But there was a flip side—I did it 24/7, month after month, by myself. Not even I can tolerate that much of me and I needed a break. I wanted to see loved ones, family and friends who I had slowly become alienated from over the last couple of years of travelling and working. Secondly, some of my equipment needed a bit of therapy; the laptop was in desperate need of a heart transplant and the bike required a decent service. I took one last look at the sign, mumbled something about 'see you in a few months' and continued on northwards in search of a decent yarn that excited me.

  My final port of call before Darwin was Daly River. I had ended up there at the insistence of a bloke I'd met the previous year on the Nullarbor. 'You'll love it', he said. 'And when you get there, be sure to ask for Joe.' Bruce was a little cryptic about it all, but he had kept in contact with me since our meeting and I trusted his judgement, which was just as well as it meant a few hundred k's more of detouring.

Northern Australia, another day another fire to ride through.

The detour was worth it as it led to a tropical retreat, an oasis forged from six months of turmoil each year. Daly River and the adjacent Aboriginal community of Nauiyu are used to the inconvenience of the 'Wet'. Each monsoon season isolates the area by swelling the river and cutting the only road in. In 1998 the floods were the biggest seen and the community had to be evacuated by air. To find out more about the flood I ended up chatting with a bloke called Dale. I figured he was in a good position to judge it, having endured 25 wet seasons as a local.

'When it was at its peak, the Water Resources people came down here and measured the flow. They reckoned that the water flowing down the Daly River was enough to fill Sydney Harbour every 7 minutes', Dale said nonchalantly.

The wet season impacts on everyone in various ways, and as Dale was self-employed he needed to do a bit of multi-skilling to ensure he had a year-round income.

'I'm a diesel mechanic by trade and I do a bit here at the community. I look after the earth moving equipment and I also grow mangoes for a living and take people on scenic river cruises', he said.

It sounded like a career-juggling nightmare. If I was in his shoes I would probably end up in a daze and start showing tourists how I could do an oil change on a mango tree. However, having multiple jobs on the go seemed to be more the rule than the exception in the Top End.

'You've got to, up here. We're seasonal. We get a lot of rain during the wet and all the earth moving equipment comes to a grinding halt. So you've got have something else to be doing because you're not getting paid sitting on your arse at home. So I'll go and tend to my mangoes. There's always something to do and you've got to be very versatile.'

However, it sounded as though most of the money making was done during the dry, and I was interested to know if that meant blokes like Dale had to work twice as hard during that period.

'Well you've got to put a few bickies away, just in case. You never know what's around the corner, so you've got to cover yourself', Dale advised.

It seemed a tenuous existence, so I asked, 'Is it a bit of a lottery living up here, in terms of not knowing what is going to blow in next week and wipe out half your business?'

> Everything was lush, healthy and chock full of chlorophyll. I was struggling to imagine how it could crank up much more.

Dale took my bluntness on the chin, gave an extended 'Yeah well' to gather his thoughts before answering.

'With my property I've sort of designed it around the flooding. The community is starting to do the same and so are the other businesses. Everything is built up above the flood level and we are prepared. So if we do get a flood, we haven't damaged all our equipment… it's just a quick clean up and business as usual once the water goes down.'

Being a southerner I had never experienced a wet season in the outback. (I once got trapped in Darwin for a few weeks during one, but that hardly counted.) I wanted to know if the wet was actually a peaceful time because of the isolation. Dale's lack of hesitation in his response spoke volumes.

'It's beautiful. We get cut off from tourism and the people from Darwin, so we have the place to ourselves. It's nice and quiet. We can go and do a bit of fishing. The wildlife seems to come to life, everything is green and there's plenty to see. It's beautiful.'

I said goodbye to Dale and went away to ponder his notion of 'green'. Compared to the deserts I had been in a couple of months earlier, Daly River was foliage overdose. Everything was lush, healthy and chock full of chlorophyll. I was struggling to imagine how it could crank up much more. Undoubtedly the Top End did undergo a Mr Hyde routine during the wet, but the concept of being cut off for several months was not something I thought I would deal with particularly well.

L to R: If you can't afford a bridge, buy a barge. During the wet, swollen rivers isolate the communities of Nauiyu and Daly River. The only way to get supplies in is via barge.

Grader and barge operator by day, croc spotter by night, Nauiyu local Joe.

Reynolds River on the Wangi Falls track through Litchfield National Park—murky, deep and boggy—all good reasons to walk it before riding it…all good except for the warning I had been given about there supposedly being crocs in it.

Spooky magnetic termite mounds ('magnetic' because they all face north…'spooky' because who would have thought termites are that smart).

A bald bloke who was kind of chuffed about making it all the way to Darwin without binning it in a major way.

Trying to get my head around the isolation factor I went off in search of my original reason for detouring into the area—Joe.

'When we've gotta get out, we go by boat', said Joe from beneath his black Caterpillar hat. 'We're used to it. It's like the people living on the islands, they get used to the cyclones coming in each year.'

My phobia about lack of freedom began to take hold and I wanted to know what happened if they needed supplies.

'There's a truck that comes out to us every fortnight. We've got a big barge here that carries about 10 pallets and we do about 5 or 6 trips across the waters to pick up the stores.'

It added up to a huge amount of manhandling every fortnight, but it was part of Joe's wet season job. During the dry he was responsible for driving a grader around the region and keeping the roads in order. I had met Joe down at the community works depot. He was dressed ready to attack the roads in clothes permanently soiled from wrestling with his ancient grader. Joe was one of those rare positive souls who was light of heart and laughed at everything. He had grown up in Nauiyu, was proud of his community and knew the bush surrounding it well. The previous night Joe had taken me to spot crocodiles in the river. Although we

only saw a few pairs of glowing red eyes, Joe reckoned it was full of saltwater crocs. In the morning I went back to have a look at the river in the daylight and was astounded to see people camping on the bank, only metres from the water. I wanted to know if Joe would do the same.

'Nup.' Joe burst out laughing before continuing. 'I wouldn't even dare to camp out down there. You know, we do a lot of spear fishing down there, but we are always on the look out for crocs and I wouldn't camp that close. The other day there was a dead wallaby on the side of the road, about 50 metres from the water, and the next day it was gone. All that was left was the tracks of a big croc that had come out of the river and up the road to get the wallaby.'

'What, right past where all those people are camping?' I asked dropped-jawed.

'Yep', responded Joe with a shake of his head and a slight chuckle.

If the campers were stupid enough to stay after having a bloody great croc traipse past their tents, I couldn't help but wonder if locals like Joe viewed tourists as liabilities or another form of entertainment?

'I don't know, maybe a bit of both', Joe said, laughing hard again.

'You know, there are signs up there for them, everywhere you go, warning about the crocs, "you can't swim" and so on. They are taking their own risk. It's not our responsibility. So if they want to park down the river next to the crocs, let 'em go.'

Before I left Joe to get on with his grading, I had one last question.

'You've told me to go on some back track to get up to Darwin, a track that's full of creek crossings and that hardly anyone travels. After what we've been talking about, I probably shouldn't be thinking about getting off and going for a dip in any of the creeks?'

'If you can see the bottom, you'll be fine,' reckoned Joe, and then burst out laughing again. I was a little unsure why he laughed this time, so I resolved to make quick work of any of the creeks that I came across.

Joe's route saw me skirt along the back of Litchfield National Park. The track was a Readers Digest version of Top End travelling: it compressed together every type of scenery and riding condition possible. I rode past billabongs, through thick jungle, creek crossings, six-foot tall grass and deep sand tracks. At one point the route cut through wide open flat lands and past massive termite mounds shaped like enormous fans. The fans were taller than me—grey, thin and all mysteriously aligned north–south.

After hours of wrestling with the bike on a tight track in stinking hot conditions, I eventually found myself back on the Stuart Highway and riding past a large blue sign saying 'Welcome to Darwin.' Twelve thousand k's, several months and countless interviews later, the bike and I had made it to Darwin, and on the surface, relatively damage free.

L to R: Random snaps… Lord knows why you'd want a bottle museum but someone in Merriwa NSW decided it was a good idea.

Equally confusing is why the blazes someone decided it was a good idea to stick a Dutch style windmill in the bush 400 k's from Perth.

The Lost City, Cape Crawford, Northern Territory.

CROCS AND CHOPPERS

'If you don't know what you're doing…and you go to a nest, the mother will come out…and you'll never go back to a nest again. It can be quite…frightening' Joe said in a slow measured tone.

The way Joe was describing it, a passer-by would have thought he was talking about raiding something like an eagle's nest, as opposed to seriously pissing off a 20ft saltwater crocodile by trying to dig up her eggs.

One of Joe's titles was 'Crocodile Breeder'. Like many self-employed people who live in the Top End, Joe had several businesses on the go (multiple jobs help people get around the vagaries of the wet season and ensure there's always some cash coming through the door). Raiding croc nests and hatching the eggs was just one of Joe's jobs. Another of Joe's businesses was hosting remote hunting expeditions, in which he and a client would hop into a helicopter and fly into remote bushland to hunt and live off the land for days at a time. Joe also ran a cattle-mustering business that dealt with mustering in very isolated country. A CV like this quickly evokes images of Steve Irwin meets Mick Dundee, but Joe couldn't be further from this image, and the last thing he was really interested in was media attention. I suspect I wouldn't have got past his front gate if it hadn't been for a mutual mate (a vet appropriately nick-named Dr Death) introducing us.

Although Joe's multi-tasking was impressive, it was specifically the croc business that I had come to talk to him about. The concept of raiding 'salty' nests sounded like the stuff that nightmares were made of.

'The best ones are where you see the mother there and you watch her go. The worst ones are the ones close to deep water and she comes out while you're at the nest. It's totally unpredictable and that's the thing with these crocodiles, you just don't know. You can check all you like, but she still might be half hidden and come out at any time', Joe said calmly from underneath his black broad-brimmed hat.

Joe was a tall lanky bloke, dressed in a long sleeve shirt and khaki shorts. He wore no shoes and the way he walked across prickly stony ground without flinching led me to believe the soles of his feet were made from sterner stuff than your average pair of Blunnies. However, being reasonably hardy is probably a prerequisite if you're going to take on crocs for a living. Raiding the nests was a two-person job: one digs while the other stands guard, heavily armed with an oar. (The first line of defence was a whack on the snout with the oar, the fall-back position involved an old revolver. Joe said they had never needed to use the revolver, but didn't clarify if there was a batting average that went with the oar.)

Joe had a young family and I asked if 'egg collecting season' was a tense time around home.

'I guess it is, it's a high-pressure time. But when you get to know an animal so well, and you follow certain rules, I don't consider it to be so dangerous. Probably the most dangerous part is the helicopter rides, going in on hot days with heavy loads', he said.

Rating helicopter rides as the most risky aspect of the process was not the answer I was expecting. Personally I would take my chance in a chopper any day rather than wrangle with a stroppy croc. But I was missing the point. Joe considered crocs to be relatively predictable, but he didn't have the same confidence in the countless intricacies that kept a helicopter in the air. Choppers, though, are central to the egg-collecting process.

'You've got to be very careful of helicopter time and man-hours, so when you do it you don't have much time to really think, you just work as fast as you can. You just follow certain rules so you don't have to think twice to know if a nest is viable or not. Like if there is deep water around the nest you don't even bother because it will take you too long to work out if the mother is there, and even then you might get bitten if she is in the water', Joe explained. 'Bitten' was probably not the word I would have gone for;

L to R: 'Krys the Savannah King. 8.63 metres or 28 feet 4 inches Estuarine Crocodile. Largest ever captured in the world'. These are the words from the plaque explaining this beastie, which is strange because I considered 'captured' to mean 'taken alive'. No such luck for Krys, who was bagged near Normanton by a woman with a bloody great big gun in 1957.

Joe, sans shoes, poking around in his croc breeding pens.

'mauled and eaten' were more the descriptions that came to mind when dealing with a tonne of croc.

Jumping in and out of helicopters and thieving the eggs of a dangerous animal sounded all very cavalier, but for Joe, and a select few others, it boiled down to a carefully calculated business proposition.

'First you've got to have an area to take the eggs from, then you've got to have approval from the landowner to take the eggs. You've got to have approval from Parks and Wildlife to take them. You need a market to be able to sell them and you've also got to know how to do it economically and, at the end of it, you've got to not be faint-hearted about getting the eggs from the nest', Joe explained.

Getting the eggs was the first part of the process; the next was hatching them and looking after the young crocs. Once they had grown to around 20 or so centimetres in length they would be sold to crocodile farms. Joe took me down to view the nursery pens. They were shallow concrete pools, secured by a small, galvanised fence and roofed under green PVC that gave the enclosure a steamy jungle feel. As we entered the pens all but two of the young crocs darted for shelter under a piece of tin. The couple that remained sat motionless in the middle of their pools and stared at us defiantly. Joe pointed at them and said, 'When they grow up, they are the ones that you've got to watch for—they have no fear and will do you harm'.

Satellite classrooms

A day after arriving in Darwin I was gazing out the window of a plane, staring down at the red earth I had been crawling across for the last few months. Flying home for a break, and having the bike trucked down after me, was a concept I had struggled with. I didn't like the idea of breaking up a journey in such a major way, but I desperately needed the rest. The Bloke on the Bike had so far been part of 1 1/2 years of lunacy: if I wasn't planning or preparing, I was on the road. I needed some time off before tackling the final section of bush bashing.

The next part of the journey was going to cover the eastern half of the country. I was looking forward to exploring the Birdsville Track and continuing on into the Gulf country, but I had reservations about the route planned beyond the Gulf. The journey would involve crisscrossing my way

south down the Great Dividing Range, as the ABC wanted me to visit some radio stations along the east coast. The problem was that this wasn't exactly the outback. I resolved to worry about that later and concentrated on looking out the window. The speed the plane was travelling made a mockery of the way I had struggled across the landscape beneath us, but as I gazed below I couldn't help but wonder who and what I was missing meeting and seeing on the tracks below. I settled back into my seat and worked on overcoming the thought that I was cheating by flying. This wasn't hard to do, as I knew I would soon be back down there. The plane was just a way to get to a destination—it had nothing to do with travelling.

The time back at base evaporated. Before I knew it I was again on the road, terrorising people with my microphone. I was at the bottom of the Birdsville Track and victim for the day was a 10-year-old School of the Air student called Stephen. I stood in front of him listening to his words but my mind was elsewhere—off in the same vacant space it had been for the last week on the road, pondering the notion of home and where exactly that was for me now.

The transient life I was living had started to leave a mark. Being on the road now seemed the place I was most comfortable. I felt at home among strangers and at ease with the quick relationships I established while travelling, but it was the long-term stuff I was having trouble with. Unfortunately this was a familiar sensation—my previous journeys overseas had left me in the same boat. I was great at starting roadside conversations but crap at dealing with the same bunch of mates 3 weekends in a row. I seemed to run

> The plane was just a way to get to a destination—it had nothing to do with travelling.

231

out of conversation; in fact I was beginning to wonder if the only conversation I was capable of was interviewing people. I suspected that too much time on my own was the problem. Maybe sensory overload also played a part—my regular diet of meeting and seeing the unusual was making living back in the real world difficult to adjust to. I took comfort in

the thought that I didn't have to deal with that for a few more months and dragged my attention back to Stephen and his description of studying in a digital classroom. In fact, I think it was the words 'exercise books' that snapped me out of my dream.

'Whaddaya mean you still use exercise books?' I stammered at Stephen.

I'd made a 100 k detour to visit Stephen who, I'd been told, had traded his books and School of the Air wireless for the Internet. Thanks to a new satellite connection, Stephen was now hooked up to the World Wide Web and using it for school—the devil in the detail was that he was only using it to do one lesson a week.

I shouldn't have been all that surprised about technology giving me yet another bum steer; after all, it had just delayed me a few extra days trying to get out of Adelaide. The laptop heart transplant didn't go well—a planned few hours of swapping out a dodgy hard drive resulted in two days of computer torment. The bike also spent a couple of days with the mechanics getting a bit of TLC, but I was a little miffed about the comment, 'What have you been doing to this? It was brand new last time we saw it.'

> I felt at home among strangers and at ease with the quick relationships I established while travelling, but it was the long-term stuff I was having trouble with.

With the bike, the computer and myself all semi-convalesced we made a beeline for the Flinders Ranges, Leigh Creek and then on to the Birdsville Track.

Stephen was living on a station about 50 k's out of Marree in northern South

L to R: In Northern South Australia the School of the Air was taking its first tentative steps to become School of the Internet. Stephen and his other grade 5 classmates dotted around the state were the guinea pigs.

Wilpena Pound from Bunyeroo Lookout in The Flinders Ranges. Stunning views, wicked riding.

A small snippet out of 5400 kilometres of the longest fence in the world.

Australia. Besides being located in the hottest part of South Australia, the station is also renowned for:

- having Lake Eyre in its backyard and housing the Bluebird while Donald Campbell was sorting out his land speed attempt,
- straddling the dog fence,
- having the legendary Birdsville track at the end of its driveway (granted it's a 50 k-long driveway, but the Birdsville is at the end nonetheless).

These were all good motives to visit the station, but the reason I went there was to meet Stephen. I'd heard the School of the Air in South Australia was switching to the Internet and I wanted to find out what that meant to a student using it in the outback. 'Not a lot' seemed to be the answer.

Stephen, and the rest of his Grade 5 class, were part of the first School of the Air Internet trial program. As it was a trial, only one subject was being done via the Internet, the rest was still being done via the wireless. I was a little disappointed it wasn't in full swing as I'd had a chat to a bloke in Derby, northern WA about what the Internet would mean for the School of the Air and he was predicting great changes for the students.

Even though Stephen wasn't using it all the time, he still gave me a run-through on how it all worked. The first big thing was that thanks to a web cam we could finally see what his teacher and his classmates looked like. The downside to all this was the teacher could also see them, so it severely restricted the opportunity for goofing about. The computer also gave students the chance to see each other's work. Stephen and his classmates' current Internet assignment was to design a new eggcup holder. The designs were drawn on the computer and everyone got to comment on each other's work. When the plans were finalised they had to build their holder and show it to the class the next day. (In Stephen's case 'the building' part involved convincing one of the station hands to drag out the welder, angle grinder and follow his design.)

The rest of Stephen's school work was still done the traditional way, using the wireless to chat with a teacher and then sitting down with text book and doing the practical work. All of this was done under the guidance and tutelage of his mum, Cindy. The input required from parents is the thing which surprises most people about the School of the Air—mum and dad become defacto teachers, tutoring their kids through each day's work. I imagined that the Internet might change this but Cindy said that hadn't been the case so far and she couldn't see it happening in the future.

Stephen's desk was swamped with books and exercise sheets, some of them due to be put into the post and sent off for marking. I thought I should let him and his mum get back on with the work, so I said goodbye and set off out of the station.

Random snaps from Lap 1 L to R: Esperance, paradise found
A little bit more paradise, this time about 3000 ks away at Yellow Waters in Kakadu.
Hughenden and Richmond in QLD, if you are into fossils this is the Holy Grail in Oz…I was lucky enough to spend some time helping a palaeontologist dig up bits of an Ichthyosaur.

ACCESSING THE LIBRARIES OF HARVARD

'The kids don't spend 6 hours a day on the radio. It's only a short amount of time they spend talking to their school based teacher here in Derby. Most of their learning takes place over print based correspondence, and it's usually implemented, taught, by their mother', Ron said.

Derby School of the Air in action. (Finding myself in what were essentially radio studios made me feel like I was back at work…which understandably was a little stressful so my visit was kept brief!)

Ron was a clear-spoken fellow, dressed in the standard north-west uniform of shorts and a T-shirt. Admittedly he had spruced it up by swapping the T-shirt for a polo shirt and the thongs for shoes and socks, but maybe that was because his job as headmaster of the Kimberley School of the Air required him to go that extra yard. I visited Ron at the Derby based school and he chatted to me about how the school operated. I was a little surprised to learn that the distance education model they used had been around since 1918. Ron said that over time it had been augmented with changes in technology, and the biggest revolution was just about to occur.

'When I was first involved with this school, 9 years ago, it would have been roughly 40% of our families that didn't have a telephone, let alone a fax machine or the Internet. Virtually all of our families now have a telephone and a fax machine. Half of our families have access to the Internet, and that number's growing rapidly. In 12 to 24 months, all of our families will have access to the sort of data technology that people in the city just take for granted', Ron explained.

His relaxed expression and manner belied the excitement he harboured about what the communications revolution meant for the students in his 450 000 square k classroom.

'It can potentially change it very, very significantly. It won't be too long until our HF radio lessons cease and that will be replaced by two-way data exchange. That will involve clear voice communication, our families having access to 24 hours a day Internet and, in due course, it won't be too far until we have access to video streaming and conferencing. This is how technology can really bridge the gap, bridge distance, and that was the driving force behind our schools—for people in remote locations to have access to equity in education for their kids', Ron told me.

I wondered if the communications revolution would help take some of the pressure off the families, as it is usually the mothers who do the teaching at home.

'We're in the business of kids learning, but our avenue, in terms of the kids actually succeeding, is achieved through somebody else ... through a home tutor, and more often than not that's a mum.' After a short pause, he continued: '...a mum who might be trying to teach 2 or 3 kids, perhaps might have a toddler to look after, might be working out on stock camps cooking for large teams of men. The mums are involved in running a pastoral company, they are the nurses on the cattle stations when someone falls off a horse and breaks an arm. The women we are trying to help support, through the process of teaching their kids, are really quite amazing individuals and have got a really difficult task.'

The use of the Internet for communications, and as a learning delivery mechanism, meant that there was going to be a big change in the way students learned at home.

'The capability for us to do far more direct teacher facilitation of learning becomes immense, because the capability will be there for teachers to pop in and out to families and families to pop in and out to us. Our kids will be able to access the libraries of Harvard, rather than the small school library we have travelling around in boxes in buses. The capability to access information is huge. Probably more importantly, it's the capability for our kids to access each other and their teachers and do things collaboratively, which are the hallmarks of very successful economies, very successful cultures—people being able to network and communicate. And that's what the future holds for us.'

Refrigerated fences

To get back on the Birdsville Track, Stephen's dad told me to take a short cut along a station track. The route cut through flat stony land that was painfully dry and coated with the sparse pale stubble of grass long since dead. I meandered through this emptiness and towards an endless and uninterrupted horizon—except for the presence of a fridge. A fridge dumped out in the middle of nowhere is perhaps not all that exciting, but one with a solar panel perched on top of it adds another dimension to the sight. Electric cables poking out of gaping holes drilled into the side of it quickly negated any fantasy about cold beer hiding behind the door and I had a bit of an inkling about what was inside thanks to a chat I'd had with Stephen's dad. Our discussion had been about the Dog Fence, and how parts of the 5400 k fence were going high tech and being electrified. The fence was designed to

In 50 000 k's of riding around Australia I've had two riding partners. Stephen was one, albeit briefly, as he showed me the way out of Muloorina Homestead.

keep livestock south of it safe from hungry dingoes north of it. The theory is good, but the problem is that it cops a caning in the process. Wombats try to tunnel under it, camels bash through it, horses trample over it, and so on. The ongoing assault means repairing the fence is a job for life. In a quest to curb some of the damage, assistance was sought from electricity. The theory was simple—a decent zap on the nose tends to slow most critters down. The concept of an electric fence is all very good, but in the outback there is the minor inconvenience of a lack of power points—the misplaced whitegoods were part of the solution in overcoming that.

The fridge provided a home to a couple of batteries that powered the fence and the solar cells on top kept them charged up. The solution was neat and simple, until it broke down. But the same could be said about the regular Dog Fence: it was good until a camel walked through it.

After taking a few 'fridge in the middle of nowhere' type photos, I hopped back on the bike. The track turned eastwards, slowly climbing to the top of a small ridge. From here, down below, way in the distance, I could make out the glaring white line of the Birdsville Track. I rode down the ridge and worked my way across to the brilliance that sliced its way through this mottled red-brown land. The Track was capped with white clay, harsh to stare at in the midday sun but rock hard and like a highway to ride on. The condition of the Track was in fact dangerously good, as I found out when I let the bike creep up to the 140 mark, a speed that was way too fast to deal calmly with the huge sandy holes that appeared now and then in the road. Besides the occasional scare, the track was a bit of a

> Wombats try to tunnel under it, camels bash through it, horses trample over it, and so on.

L to R: A bit of scrub, a few motley sheep and the thin white ribbon that is the Birdsville Track.

Sadly it was not a solar powered fridge packed full of cold beer for passing desert travellers. It was home to a couple of truck batteries that were pumping current through a new section of 'electrified' Dog Fence.

disappointment in terms of being the challenge I expected. I was silly enough to mention this later to a bloke who owned a service station in Birdsville and he soon set me in my place.

'We are pretty bloody happy about the condition it's in, as we have to live with it. It's not a road that has been put there for adventure tours, it's a vital link for us. The biggest purchase for a young Australian couple is a house, but out here, because of the lousy roads, they have to find 50 grand for a decent 4WD before they can even think of a house, so don't start whingeing about the good state of the track', he lectured.

That rebuke was a few days ahead of me—I still had over 500 k's of far northern South Australia to work my way through before I got to Birdsville. As I continued on up the track, the dry barren land managed somehow to become even more parched and denuded the further I rode. The notion that people actually lived out here and ran livestock seemed incomprehensible—it was the harshest and most unforgiving land I had seen in two laps of Australia. Admittedly I was seeing it in the midst of a long and terrible drought, but it was difficult to imagine it ever looking much different. The only outbreaks of vegetation were around bores or dry creek beds, but even the grey dusty growth spread sparsely around the banks of the biggest waterway of the red zone, the Cooper Creek, seemed to be only just hanging in there. The crowns of all the trees were dying back, making a slow retreat in a losing battle against the desert. I finished winding my way through the dry Cooper crossing and focused on pushing towards Mungerannie Hotel. And, as usual, I arrived in the dark.

THE DOG FENCE RIDER

'I was a dog fence rider for 20 years, patrolling a 345-kilometre section of fence. I was one of the first contractors in South Australia because most of the Dog Fence was patrolled by the landowners', John explained to me.

John had recently retired from being one of the contractors who maintained the longest fence in the world—5400 k's long. The fence stretches from Ceduna in SA to central Queensland and protects sheep on one side from hungry dingoes on the other.

Me, being an ignorant beggar in many areas and especially so when it comes to the eating habits of dingoes, was particularly surprised to hear that dingoes were such an issue.

'Oh yeah, a real big issue', John said emphatically. 'A lot of times a dingo will get into a mob of sheep and he won't just kill one and eat it, he will go through them and maul and maim them. You end up with 20 or 30 injured sheep, and there is not much you can do with them once they've been maimed.'

Admittedly the dingoes were there first and were just doing what they had always done, although natural processes didn't hold much weight with pastoralists. They decided 100 years ago to do something about the carnage, and so began building the fence. The age of the barrier alone would warrant regular maintenance, but that is not the main reason for its state of disrepair.

'Wombats dig holes under it, foxes go through it, camels knock it down, so it's a constant problem' John told me. He reckoned that camels had been a lot worse lately because of the drought—they were punching holes in the fence every 300 metres to try and get access to water in the south. It all added up to lots of big holes that required lots of work.

Most of the fence passes through remote and isolated country. The section John had patrolled was no different. Because of the remoteness, and the amount of work involved, he would go out patrolling for a week at a time and hardly ever meet anyone.

'Now and again you might see a traveller, but you don't see many people out there. It gets pretty lonely at times, but you've got to be used to your own company and make do with your own company. Plus you've got to put up with the heat and the flies and you've really got to like the bush. If you don't like the bush you won't be out there, you just won't last', John said.

Occasionally when John did see travellers they were a good source of entertainment.

'There were 3 German tourists in a Land Cruiser and they hadn't got through to Tarcoola—they were supposed to contact me when they got through. So the Police went out to search for them. Eventually they were found, bogged up to their axles in the sand and nearly out of tucker and water. One of the policemen went up to their 4WD and walked around it to check it out. He stopped at the front wheels, bent down and locked the hubs in to engage 4WD, then hopped in and simply drove it out of the sand. This lot had been sitting there bogged for 6 days because they hadn't engaged 4WD properly', laughed John.

It seemed that dopey travellers were probably one of the constants, as even the fence was undergoing some change. After nearly 100 years of using wire netting there was a shift in construction, and it was all thanks to solar panels.

'There's a lot of electric fence going up now, because it's just too dear to replace the old netting fence, so they've gone to a 10-wire electric fence', John explained.

As usual I couldn't hold my tongue and before thinking I blurted out, 'So you've got to be an electrician as well?'

'You've got to understand solar panels, energisers, batteries—it's all a new ball game compared to what it used to be', John replied.

Besides reducing the maintenance costs, it also sounded like the new electric fence was less of a barrier for other animals. John said roos could now jump straight over the top of the new fence as it was only about a metre high. One of the problems the fence sometimes causes is an imbalance with wildlife;

kangaroos and wombats don't have predators south of the fence and the result is they sometimes end up in plague proportions.

'I've seen dingoes go wombats', John told me. 'I saw this wombat that dug a hole only big enough to get its head in because it was trying to get away from a dingo. So I shot the dingo and when I pulled the wombat out it had both its ears chewed off, so the dingo had already been at him. The dingoes definitely clean up the wombats and there's more wombats on the inside of the fence than what there is outside, because of the dingoes.'

'I'm heading west—any advice? Like, don't hit a wombat?' I asked.

'Oh yeah, definitely don't hit a wombat! It's like hitting a brick wall', John said laughing. 'We might not see you again if you hit a wombat. They are tough little buggers, solid. Good eating though, just like pork. They are like pigs', John added, a little unexpectedly.

I looked at him somewhat surprised and said, 'Well if I hit one mate, I'm not sure if I'll get much chance to cook it and try out your recommendation!' And laughing, I rode off.

⬢28

Five of the seven most deadly

'We've come up from Sydney to drop some golf clubs off', Dave said.

'So several days, and a few thousand k's just to drop off a set of golf clubs?' I asked, a little stunned.

'It's probably about a 3-day trip', Dave replied. 'But, as you do in these parts, we decided to stop off and check out the hospitality at a few of these places', Dave said and smiled broadly.

A bloke telling you he's driven a few thousand k's just to drop off a set of golf clubs is probably the sort of yarn you would expect to hear in a pub

Terry, Dave and Johnno. The blokes on the left had just driven from Sydney to deliver some golf clubs to Mungerannie Hotel on the Birdsville Track. Johnno, Mungerannie's manager, was particularly pleased as it meant more people could get on the course—which included a mandatory 'bar hole' each round.

in the middle of the desert. The difference with Dave was that he was telling the truth. Mungerannie Hotel, besides supplying fuel, beer, a bar, accommodation and food, had recently added golf to its list of amenities. However, there was a problem with their most recent addition: they only had two golf clubs. Dave and his mate Terry had taken a shine to Mungerannie a few months earlier when travelling past, so when they got home they decided the decent thing to do was to donate some clubs. But finding a freight company that was happy to deliver golf clubs halfway along the Birdsville track proved difficult, so they said 'Bugger it, we'll drive back out there and deliver them ourselves'. Now Mungerannie not only featured a course but they also had clubs to use on it.

To go to the effort of shoving in a course, albeit a unique bush version of one, you would think that Johnno, the manager of Mungerannie, was a pretty keen fan of chasing the 'dotted ball'.

'No, no, I hate the game really, it's too frustrating,' Johnno said, laughing at the irony.

Johnno was a tall lanky bloke, probably better built for speed bowling than golf. If he did take to thundering down a pitch I imagine his beard would get in the way—it was long, grey and flowing. Johnno's face and beard were sheltered under a brown and beaten bushman's hat. He seemed to me to be like Bill Barnacle from *The Magic Pudding*. I tried to push Norman Lindsay's drawings out of my head and focus back on our chat about the course. I suspected that Johnno had only put in a 3-holes course so that players were never far from the bar, and I asked them if this was the case.

'It's compulsory to have a beer after every 3 holes. The third green comes right into the clubhouse, shall we say', he said laughing.

Some would argue that describing what sits out the front of the Mungerannie Hotel as a golf course is not only being loose with the description but is also putting a slur on the proud tradition of golf. The quality of the course probably has a lot to do with its 'spur of the moment' origins.

'We had a tip truck around here doing a bit of work, so we just used white clay for the greens. We dumped that down and spread it out with the loader. It needs a little bit of touching up. In fact the greens are a lot rougher than the fairway, but that all adds to the excitement. And that's why I've got a 20-litre drum for the hole, so you've got a little bit better chance of getting it in', Johnno said with a chuckle.

If golf wanted to get serious about broadening its appeal and attract more adventurous players then some tips should be taken from the Mungerannie course. It regularly has five of the most lethal snakes in the world slithering across it, in summer the temperature rarely dips below forty degrees, the holes are made of 20 litre drums and to really mess with your head the 'greens' are made from white stones that are roughly the size of golf balls.

The course looked fit for the Flintstones—it was covered in rubble and the flags marking the holes were red rags nailed to old tree trunks. I imagined that with 20-litre holes there was a reasonable chance of getting a hole in one. Johnno was quick to point out that it wasn't as easy as it looked.

'You would have to get it right in. There's not a lot of roll on those greens and you've got to get around the stones.'

The stones were pure white limestone, also about golf ball size, and the greens were also stark white because of the clay used—it was probably one of the few 'greens' in the world you could lose a golf ball in the middle of. With a pure white course glaring away under the blazing sun I was thinking Johnno probably should put up signs warning players of possible retina damage. Glare, stones, saltbush and sand were all significant handicaps to deal with. To up the ante a bit more, there are also snakes.

'Oh yeah, we've got 5 of the 7 most deadly snakes in the world, so you've got to be a little bit careful', advised Johnno.

'So it is probably a little bit past a par 11 if you taken in the danger factor?' I asked.

Johnno laughed and said that was especially the case in summer. I'd heard some nasty

> The course looked fit for the Flintstones—it was covered in rubble and the flags marking the holes were red rags nailed to old tree trunks.

245

stories about summer in this part of the world, so I was happy for our conversation to turn towards the topic of heat.

'It reaches over 50 degrees in the shade. We get a few 51s, so out on the flats it's 60, 65.'

These were not just freak days Johnno was talking about, as he went on to make clear.

'The summer before last was pretty hot, she was over 45 for about 6 weeks.'

Despite such extremes, Johnno didn't talk about summer as a trial; in fact it sounded almost as though he looked forward to it.

'It's quiet out here at that time of the year…come summer, that's our holiday time.'

With such evil temperatures, and only 4 inches of rain a year, existence would be impossible if not for the Artesian Basin. The Basin is the lifeblood for much of central eastern Australia, Mungerannie included. Not far from the roadhouse, water burst out of the basin at 125 psi, a pressure that well and truly puts the average garden tap to shame. But I guess it's not all that surprising when you consider that there is the weight of over 1000 metres of rock and soil siting on top of it. At Mungerannie, water coming up from this depth is hot—around about 85 degrees hot—which meant it needed to be held in storage tanks and allowed to cool before it could be used in the 'cold' taps around the hotel. Pressure and temperature aside, the bore spat out an enormous amount of water each day. Johnno didn't say how much but it was definitely way more than he could use. So much so that the excess from the bore created an interesting by-product—wetlands—which led to Johnno naming the place 'Mungerannie by the Sea'.

'The bore drain has been running in there for over 100 years. There is 140 species of birds down there. On the other side of the waterhole the sand dunes come right down to the water, and there is a bit of salinity in the bore water. So it has to be 'Mungerannie by the Sea' because we've got everything the sea has got—we even get pelicans and seagulls out here.'

Mungerannie was a hell of a detour for your average coastal-dwelling, chip-scabbing gull. Truth was, it was also a fair detour for Johnno and his family—they had come from central Queensland and it was a move that wasn't initially embraced by Johnno's wife Genevieve.

'He dragged me down here by the hair. I gave him 6 months and then I fell in love with Mungerannie and the locals that live along the track', she said flatly.

The Mungerannie Spa and wetlands.

'Simple as that?' I asked.

'Yep, and the people out here are legendary. All the folk along the track work really hard and then they play really hard. It's the closest-knit community I have ever lived in', Genevieve said proudly.

I was a little surprised to hear this and blurted out, 'That's almost an irony, because out here you think there is no community because you are so isolated'.

'It's a real community, the neighbours genuinely look after each other. If there is any trouble along the track the neighbours just drop everything and head to the nearest station where they are needed, either with sandwiches, physical labour or counselling, and then usually a beer or a wine or a Black Rat afterwards', she said smiling.

The Black Rat threw me, and I focused on getting to the bottom of that rather than absorbing what Genevieve had just said. Once I got it sorted that it was Track slang for 'Rum and Coke in a can', I could get back on with the job of putting my foot in my mouth by totally ignoring what Genevieve had said about the track being a community.

'Taking a step back, you talk about neighbours, how close is your nearest neighbour?' I asked laughing.

'Well, just over the sand dune at the back is Mungerannie Station', Genevieve said, nodding in a direction over my shoulder.

'People say I am very isolated, but it's only 200 metres to our nearest neighbour. The next lot are 40 kilometres away, the next lot 60 kilometres away. But all our neighbours are people that live along the Track, so we call our neighbours people who live 300 kilometres away as well.'

Thanks to the travelling I had been doing I shared a similar perception to Genevieve about distances and isolation, so I wasn't sure why I had asked such a dopey question. Maybe my brain hadn't got out of 'town distances' after my break in the city, or maybe it was because the tape recorder was running and I was trying to put it in terms that would be relevant to an audience in a city. Whatever the reason, the bottom line was I feared Genevieve was thinking I was a bit of an arse and I needed to recover.

Pre-Mungerannie, Genevieve was an occupational therapist and I was interested in how the move had changed her.

'I can now service a generator. Because I travel by myself I am fairly good at mechanics. I can change flat tyres, pull tourists out of bogs', Genevieve replied. 'I've learnt to be fairly independent from Johnno.'

Genevieve said the independence was the result of them occasionally being separated when Johnno went away or she was out on the Track by herself. Ironically, separation was the only downside Johnno listed about living out there, and it wasn't in regards to Genevieve—it was about their children. It seemed the penalty they had to pay for living at 'Mungerannie by the Sea' was that when the kids reached 12 they had to leave and go to the boarding school. Johnno tried to downplay it by jokingly saying it was a bit of a bugger because at that age they were starting to get handy and could to do a fair bit around the place.

I left Johnno to get on with his day and my mind harked back to Stephen and his satellite Internet connection. I wondered if the bold new digital world meant that in the future families in the bush would get to stay together longer, rather than be split up to meet their educational needs. It was a nice notion, but I suspected it might remain a fantasy. After all, there's a hell of a lot that you can't learn through a computer.

> I wondered if the bold new digital world meant that in the future families in the bush would get to stay together longer, rather than be split up to meet their educational needs.

29
The Commodore of Birdsville

Mungerannie was a place I was in no particular hurry to leave—it was a refuge from the hostilities of the surrounding environment. Besides the sanctuary status, there was also a steady trickle of travellers passing through, and most of them usually had something interesting to say. A final reason to stay, albeit a selfish one, was that Johnno was a bike fan. He took me out the back of the hotel and proudly opened up a shipping container—there under a sheet were two Harleys. Johnno explained that one was a 'town bike', which was always carried down to the start of the bitumen in the back of a ute; the other was his 'Track bike'. I baulked at the

A few hundred cockies goofing off on the side of the Birdsville track...I suspect the lack of a decent size tree for a few hundred k's meant Sturt Stones were the only place to roost and contemplate why the hell they were flying around a barren desert.

thought of riding a Harley on the Birdsville Track and Johnno admitted that it was a little tricky in sandy patches, but the fact that he even did it spoke volumes about the generally good condition of the Track. However it seemed I had seen the best of it, Johnno said the conditions weren't as good north of Mungerannie and warned me to stay alert.

The hotel hadn't long vanished out of my rear-view mirrors before I found out what Johnno was on about. Sections of the road became giant potholes that stretched for hundreds of metres and were full of pulverised clay and sand. Spotting the holes was a problem because the glare screaming off the white road made it difficult to see any damaged sections until I was on top of them. I was forced to slow down, travelling at a pace that ensured I could deal readily with these, which was frustrating as they were only present in a small percentage of the Track.

As I pushed northwards the baked landscape had another layer of harshness pasted over the top of it. Small, glimmering brown rocks now carpeted the scorched earth in their billions. The small pebbles had been baked for countless millennia in an oven known as Sturt's Stony Desert. Each day the stones were cooked a little more while an invisible hand buffed them shinier and smoother. The landscape showed no signs of life: all vegetation had been locked out by a mat of stones. The rocks threw up a shimmering reflection, a permanent mirage that kept me company as I rode. The only break in the shimmer came from a flock of sulphur-crested cockatoos, thousands of them sitting on the ground because there was nowhere to perch. The birds didn't stir as I

L to R: In 25 000 k's I dropped the bike three times and not only was I dopey enough to photograph each I was also stupid enough to let the publisher print them.

Sturt's Stony Desert.

approached. They remained motionless, resting, as though conserving energy while pondering how they had come to be in the middle of the desert.

To get into Birdsville, Johnno advised me to detour off the main track and do the last part on something called the 'Inside Track'. The track appeared to have been made by simply scraping Sturt's stones aside. It headed west, and the glare off the rocks was burning into my eyes as I rode into the afternoon sun. My eyes got a rest when the track was forced northwards by sandhills—6-metre high ridges heralded the edge of the Simpson Desert. The track followed the contours of the sandhills before peeling off and coursing through dried up marshland, and suddenly I was surrounded by vegetation and involved in the most challenging riding since the Flinders Ranges, nearly 1000 k's ago. The track twisted left and right, darted through trees and surged across grasslands—I was having a ball. In places it was sandy but that no longer bothered me; I had learnt that sand was largely a confidence trick. My confidence got a kick in the pants when I flew over a small ridge that was hard packed soil on the side I went up but soft sand on the obscured downhill side. Unprepared, and stupid, I let the front wheel dig in and binned the bike with a nosedive. I rolled down the ridge and picked myself up out of the sand at the bottom, uninjured. I marvelled at my good

> The track appeared to have been made by simply scraping Sturt's stones aside.

fortune on the health front and cursed my stupidity as I then trudged back up to the bike. It seemed undamaged so I set about unloading and digging it out. I was working quickly in case another vehicle came roaring over the ridge—I was completely obscured and would have made a bugger of a speed hump. I was working on the premise that I had not seen another vehicle for hours, so now would be the time for one to turn up.

Extracted and reloaded I rode off. Soon I trundled over what I figured to be the Queensland–South Australia border. An ancient wooden fence was the only tip that it might be a border, even the GPS didn't seem too clear on the issue.

A short ride later, and in ailing light, I could make out the outline of a town—Birdsville.

A distraction was needed. It came in the form of a barking dog yapping excitedly from the back of a small truck. The dog was probably overjoyed its teeth were still in its head, the truck had just finished rattling over the millions of corrugations that have to be endured before reaching the smooth tarred oasis of Birdsville. The driver of the truck, however, did not appear to share his dog's enthusiasm for the bitumen. The truck ground to a halt, a whip-wielding arm shot out the window and flicked backwards, the air above the dog's head splitting with a loud crack. The dog got the message and the truck was off again, well before the cloud of dust that heralds every arrival into town had time to catch up.

Dog control administered by a whip, blindly and backwards from the comfort of a driver's seat, would usually be something I would view with a degree of a surprise. Considering, however, that I was in the middle of a desert, listening to a bloke tell me he was soon to be the commodore of the local yacht squadron, it was going to take more than Birdsville's answer to Dr Harry to seriously distract my attention.

The 'dog and whip' sideshow did, however, give me some precious seconds to gather my composure and refocus on the task at hand. The interview I was doing with soon-to-be 'Commodore Wolfgang' had got off to a bad start; it seemed my microphone had not been plugged in for the first 5 minutes of the recording. Just as I was admitting to my incompetence the dog arrived on the scene, giving me a few precious seconds to try and re-establish the charade of looking like I knew what I was doing. Building credibility was an uphill battle, as I feared Wolfgang already had questions over my state of mind because I was doing the interview in my riding suit—not exactly comfort wear for 30+ degrees. However, any misgivings he may have had remained hidden, despite the microphone debacle.

Wolfgang—artist and future commodore of the Birdsville Yacht Squadron.

Wolfgang did not immediately strike me as 'commodore' material (the absence of any form of a harbour for over 1000 k's probably had a lot to do with this). Wolfgang was somewhere between 30 and 60 years old; the tinge of grey in the beard suggested more towards the upper end of the scale, but the unkempt thick black mane and his ability to laugh at 'the ludicrous' pointed towards the bottom of the scale. However, when he said he left Germany because he figured the same people who had caused the war and ruined his life were still in charge, I got a better handle on his true age.

Birdsville is not the easiest place to get to—it had taken me about 40 000 k's, a quick change in motorcycles and a couple of years. For Wolfgang it had taken most of his life, a vast fortune and a couple of wives.

'I fell in love with the extremes of nature—have always been in love with actually—the sea, the mountains, the desert. I first got out here around 1976 and discovered what Australia was really about. I call these people who live on

> Birdsville is not the easiest place to get to—it had taken me about 40 000 k's, a quick change in motorcycles and about a couple of years. For Wolfgang it had taken most of his life, a vast fortune and a couple of wives.

253

the coast, where everybody loves to live, "the fringe dwellers"', cackled Wolfgang before continuing.

'With modern facilities you can be anywhere at any time. I always say to people that we are centrally located in Birdsville—we're equally far from anywhere and effectively only 30 hours from Paris.'

I laughed, but Wolfgang didn't. Either he had said it so many times that it was no longer amusing or I failed to appreciate the point he was trying to make because my judgement was slowly being cooked inside the riding suit.

Realising my gaff I went about digging myself in deeper.

'But you're an artist, you paint the desert, don't you need to be near the galleries to push your product?'

'They say 60 000 people travel through here each year. I doubt that very much. However, it's the travellers who buy the paintings, they don't mind spending money. They spent $120 000 on their 4WD, so they don't mind spending $1000 on a painting', Wolfgang said.

I nodded and tried to look like I was thinking, 'Struth, if that's all they cost I might just ditch the camera and buy paintings from here on in'.

Wolfgang ignored my brow-knitting and continued.

'But I do earn my money in painting. It was sort of a development. It ran parallel with my discovery of the desert, my understanding of the desert. It's a progressive thing—you just don't come out here and think you know everything. I painted the desert landscape for many, many years, since the 70s. But to understand this country I think you have to drown yourself in it. You have to see the seasonal changes and the people with it. That's a process that affected me and it's still doing it.'

Seasonal changes are what enabled Wolfgang to pursue his other great passion—sailing. Birdsville and the surrounding Simpson and Sturt Deserts are not renowned for their rainfall. They do, however, get floods, usually monstrous ones from monsoon rains up north.

'We had two floods in 2000, which were 7 metres above the norm. The Diamantina, in places, was 60 kilometres wide and we are totally isolated. The year before, the Eyre Creek flooded and we are totally encircled, the only way in is by plane. Our supplies come up by transport to Pandi Pandi and 50 kilometres down the river we go, by boat, to collect our supplies. It looks like the Amazon, you've got dingoes and snakes and lizards swimming across the river and the bird life is just phenomenal,' Wolfgang reminisced.

From where Wolfgang and I sat it was less than a stone's throw to the edge of town. Beyond this lay endless tracts of sunburnt rocks and scorched soil. The heat radiating from them repelled my efforts to imagine it submerged. But the floods do come, and they are responsible for Wolfgang's commodore aspirations. He is probably the only person living in the desert who owns 4 boats and a windsurfer. One of these is a sailing boat moored on a permanent waterhole called Andrewilla, 75 k's south of the town.

'The waterhole is about a 120 metres wide and 11 kilometres long. There is nothing like running in front of a sandstorm, between the two sand dunes. It's like sailing on the Suez canal!' He laughed at the comparison before moving on.

'I want to form a yacht squadron here in Birdsville and hold a regatta on the waterhole. I know the commodore of the Middle Harbour Yacht Club in Sydney and he comes here frequently. We are thinking of having a Laser competition, which is a dinghy. We know the Laser champion of the world, and he is going to come up and have a proper course set up and there will be serious racing here in Birdsville. I will then form the yacht squadron and I will be commodore', chuckled Wolfgang.

'So it's exactly what you'd expect to find in the outback. A German-born artist, commodore of his own local yacht squadron, running a race between two sand dunes', I said.

'Yah, it is very typical I would say. You thought it was all Ringers and Cowboys!'

POSTSCRIPT My travels did not finish in Birdsville—there was another 10 000 k's of riding before I got home. But in some sense it was the end of the 'journey', as little of the remaining ride was through the outback. From Birdsville I pushed on to the Gulf of Carpenteria and rode west to east along the bottom of its huge expanse. The Gulf Track had remote communities dotted along it and was traversed by people ranging from travelling sideshow operators to university students roaming the top end to survey fresh swordfish numbers. (Yep, believe it or not, such things exist in our northern waterways and grow up to 5 metres in length!) The flavour of the journey, however, had begun to change. Compared to other outback areas I had travelled there were a lot more people and traffic (granted it wasn't Bourke St, but 20 cars a day seemed like peak hour). The further east I went the busier it got. Because of ABC commitments on the coast I had to go, but considering this is where most Australians live it is kind of hard to call it the outback.

In some regards I think I also became a little warped (well more than normal). If I wasn't rocking into a town with more than vapours in my petrol tank and a mouthful left in my waterbag, then I was having it easy and couldn't really consider myself to be in the outback. This reckless attitude was partly the result of being laden down with technical equipment and not having room for additional stores of water and petrol, let alone any food. Having said that, it was also a calculated risk; the problem was that the further I went, the less calculations I bothered to do. Perhaps the reason for this was the further I travelled the less remote I considered myself to be, and my definition of what was outback became more and more extreme. Maybe my perception was also being coloured by the satellite dishes and phone towers that were sprouting up across the bush. It was stupid really—just because a place was now wired up to the outside world that didn't overcome the fact that it was still a 3-day drive to see a doctor or dentist. I suppose the difference was that now they could phone up and say, 'We are coming and hopefully the 2000 kilometres of corrugations won't rattle out all our teeth before we get there!' My faith in technology and communications also explained some of the risks I took. A satellite phone, a personal EPIRB and a GPS gave me a safety net that didn't exist 10 years ago (admittedly it was a fat lot of good if I ended up lying unconscious on an obscure track that no one travelled).

Along the Gulf Track I met a Brit who was busy circumnavigating the planet in a Land Rover. From England, he had driven up to the Arctic Circle and then made his way down through Asia to Australia. He had a lot of k's under his belt and he summed up the advantage the outback had over places he had travelled in the world.

'You've got the remoteness, but at the same time it's nice to know you've got the backup if something goes wrong. You've got a fantastic backup infrastructure, god forbid you ever have to use it though.'

Travelling alone on a bike … maybe I'd been lucky to not have to use the backup, I don't know. My luck aside, I do know it gave me the confidence to get out to people and places that, sadly, few others do. But that too is changing—caravans trundling down the Birdsville track is testament to that. Purists might argue that if you can tow a caravan into a remote location and use your mobile when you get there then it's not really the outback. In the end, though, the definitions are a load of bollocks, because it's about the journey, not the destination.

Sprechen sie what?

My language occasionally caused confusion among some of those that helped put the book together. It was suggested that some clear explanations of a few of the colloquialisms might be a good plan…so here we go.

LIST OF COLLOQUIALISMS

amped	pumped, stoked, buzzed	hardcore	full on, hard yards
balls up	dogs breakfast, cock up, screw up, bloody shambles	k, k's	click, metric mile, French mile
		lairy	loud, yobbish
barra	scaly bugger that blokes in Northern Australia tell lots of lies about	loon	boofhead, dope, drongo, Galah, half-wit, mug, noddy, thick head, twit
bar research	elbow work, sink a few, give it a nudge	manky	mangy, wiffy, on the nose
		metric tonne	100 k's per hour, not fast enough!
beans	as in 'give it the beans', give it some welly, feed it a fistful	mickey mouse	see 'ducks'
bed flute	trouser snake, pyjama python, love wand, Mr Happy, Chester, ferret	no wukkas	no sweat, no probs
		ockie strap	God's gift to travelling motorcyclists
bee's dick	two fifths of five eights of stuff all, bugger all	pear-shaped	turned to mud, went to crap, arse up
binning	as in 'bin it', stack, bingle, a big off	pushies	pushy
		reccy	suss out, perve, optic nerve
blag	as in tell a few minor furphies, rort, snow job, wangle, gyp, diddle	redlining	maximum beans!
		rubbernecking	sussing out, having a perv/gander
blokey	blokey	scone	nut, noggin
blouse	wuss, soft option, lame	six-pack	brick, quarter slab
blunnies	foot clobber	speccy	speccy
bollocks	as 'in the dogs' bollocks', ducks guts, grouse, tops	suss out	see reccy
		threads	duds, clobber
cactus	carked it, snaffooed, rooted, clapped out, on the fritz, had the dick	U-bolt	180
		upping stumps	chuck in the towel, call it quits
		up the ante	turn it up, razz up
date	ring, quoit, where your mother never kissed	vego	herbivore
		whooped out	tsunami sized corrugations
digs	pad, shack, place, joint	wizza	slash, bleed the lizard, point Percy, splash the boots
ducks	see bollocks		
duds	daks, strides, britches	wuss	see blouse
fanging	hooning		
goolies	nads, nuts, Niagara Falls, spuds, orchestra stalls		

Tech stuff

BIKE CHOICE AND PREPARATION

The following is the only psuedo bike techo section in the book. I've shoved it in for those who have a similar sad bent and also because I can finally answer a lot of questions about the bike.

CHOOSING SOME WHEELS

Where do you start and where do you stop in the quest for the perfect bike set up? Unpacking a purpose-built bike, fresh out of the crate, is the simple answer. Long-distance bike riders usually have to spread thin budgets a long way i.e. buying a bike, setting it and then covering all of the travelling expenses. For many bike travellers, a brand new set of wheels are really only possible when Saturday night's lotto numbers are known in advance. My lack of 'lotto prescience' meant there was only one other option available: cheat. Perhaps, more accurately, I was spoilt. Each time I headed off around the country the ABC leased me a ride. This meant not only did I benefit from the trouble free riding a new bike brings, it also meant I could choose the perfect bike for the job—a BMW R1150GS.

I feel like I have just uttered some blasphemy by saying that, but let me explain. One of the rules of the Government Broadcaster is that any form of endorsement is absolutely verboten, which also includes the use of brand names. So for a couple of years I've really only been able to refer to the bike as 'The Bavarian' or 'The Bike'. Rabbiting on about performance and features was also an absolute no-no. Finally, thanks to the book, I can now say a few things about the bike.

If ever there was a beastie built for what I was going to do, it was the dual-purpose (on road-off road) BMW GS bikes. Hopefully I won't wake up with a horse's head in my bed after saying that—bike riders can be fragile creatures, they continually seek reassurance that their set of wheels is the 'ducks'. Suggestions to the contrary don't always go down that well. The following is my list of reasons for choosing these bikes.

Anti Lock Brakes (ABS) My prime directive for the two years on the road had nothing to do with gathering stories; it was my boss saying 'just don't fall off the bloody bike'. The GS was the only bike in this class of bikes that offered ABS. The feature made my boss happy and I was kind of chuffed about it as well.

Shaft drive I had dealt with the joys of chains and sprockets on other bikes, including when I rode from England to Australia. I was over the tightening, oiling, mess, replacing and so on involved with a chain. Considering the k's I would be doing and the conditions I would be riding through, a chain was going to be a heap of pain that I just did not want to sign up for. Again, the GS was the only bike with this feature.

Comfortable When I was on the road the standard opening greeting from people was, 'So how is your bum?' To appease them, I would joke about being on a waiting list for some butt cheek transplants. In reality, it wasn't an issue. I've had to do a few 14-hour days in the saddle, and although they were absolute nightmare rides I never had any pain as a result of them.

Rated to carry a lot of weight This was essential for the ridiculous amount of equipment I was carrying.

Factory supplied luggage and engine protection bars The hassles I've always had with bikes has been with the after-market products bolted on to them. It was because of such items that I had the joy of finding welders everywhere between England and Australia (which is why I called my first series 'Asian Welding Tales'). The GS stuff is factory designed and it fits and works. The luggage doesn't require any poxy straps, just clip, lock and go.

Heated hand grips Very happy about this while riding past snowdrifts in Tasmania and the Australian Alps.

10 000 k service interval With Australia being so ridiculously big this was very handy. However, old habits died hard and I still dropped the oil every 5000 k's.

Reliability All of the above adds up to a reliable bike and this is something I am very keen on. I've done my time tinkering with bikes—it's overrated. The only thing I had to worry about was keeping the petrol tank full, and there is even a fuel gauge and warning light to help keep this under control. (Essentially this bike was Drew-proof – which is saying something.)

BMW lease BMW seemed to be the only people who would lease us a bike.

BMW roadside assist If it did go pear-shaped, no matter where, BMW would sort the problem out—I never had to test out the service but it did give peace of mind that I would have killed for in the deserts of Pakistan.

POST CHOICE

With Bike #1 I did very little other than run it in and go. The only additional thing I did was clip on an after-market top box. This was custom fitted with foam, like a camera case, so it would protect the laptop, camera etc. This box unfortunately re-acquainted me with my fears about after-market products by duly breaking off and bouncing $15 000 worth of gear along a dusty track. Sadly I had no other choice than to use this box, as it was the only one big enough to carry all my gear—it failed on both journeys!

A QUICK CHANGE IN BIKES

Journey 1 was on the bitumen the bulk of the time and was less demanding on the bike. Journey 2 was a different ball game; I was heading bush and I needed to make some modifications to get the bike up to battle spec. I contacted Justin at BMW and gave him my shopping list. His response was, 'We can do better than modify, we've got a new model that will suit your needs'. There was, however, one minor catch in swapping bikes—the new model wasn't available until two weeks before I was due to depart.

The wait for the 'Adventure' model was worth it. The modifications included:

Lower first gear Makes navigating rocky river crossings and tight sandy tracks a whole lot easier.

Increased suspension travel Better at soaking up wash-outs and truck-eating corrugations.

Coded engine plug In remote areas the fuel quality is often beyond abysmal and causes some nasty engine knock. The coded plug clips into the engine management system and effectively detunes the bike so it can cope better with lower grade fuels.

New luggage system This meant extra space so I could carry even more equipment. I needed this as I was shooting my own TV stories on journey 2.

Knobby tyres Makes dealing with countless k's of dirt and sand a whole lot more predictable. Unfortunately, though, the power and mass of the bike leads to a short tyre life—6000 k's for a rear, if I was lucky and very careful and didn't mind riding on canvas at the end.

Braided steel brake lines Gives more feel when braking, and especially helpful if you need to grab a big handful of front brake without the wheel locking up and sending you chin first down the road. (Yes, ABS brakes should avoid this. However, using ABS on the dirt is a bad plan. Being able to lock up the rear wheel is essential for dirt riding.)

Two built-in power points Useful for running a GPS and other devices.

30 litre tank Vital for covering some of the long hauls between top-ups.

It all added up to the perfect bike for my needs; the only downside with all these features is that it ends up being a massive bike, which, to be perfectly honest, freaked me a little when I first got on it. The thought of hurtling down a dirt road on it required a leap of faith. The trick I soon learnt was to forget about the size and ride it like a big trail bike and it will do just about anything. It's a joy to ride, totally predictable and has probably saved my bacon more times than I care to count.

PRE-FLIGHT PREPARATION

With the bike sorted, the rest of the tinkering was largely about setting up my living space by adjusting the suspension, levers, handlebars and working out how to pack everything etc. Having said that, there were a few other things to consider:

Security With all the technical gear I was carrying, a halfwit with a jemmy bar and a spare 20 seconds could easily have scored themselves thousands of dollars worth of goodies. An alarm was vital, but unfortunately I had to go after-market. Again my fears of 'non-genuine' were reinforced when the bracket holding the siren chewed through the fuel line when I was in the middle of nowhere.

Throttle lock Sounds dangerous and potentially is, but it's fantastic for giving the throttle hand a bit of rest. (Also a lot of fun because it lets you ride no hands standing up and scare the blazes out of passing motorists in the middle of nowhere —just joking kids.)

Diff snorkel The diff has a breather, which unfortunately can suck water when submerged. Since I am rather fond of using bikes as U-Boats, a snorkel pipe was fitted to make sure there was no mixing of oil and water. Regrettably I did not find out about the benefits of a snorkel until after I had blown a diff seal.

Charging systems Access to 240v was often days apart, so many an hour was spent setting up systems for recharging the laptop and camera while riding along.

Contingency plans—zip ties and gaff tape—there isn't a problem that can't be fixed with a fistful of these.

WEAR AND TEAR

If you spend a couple of years and 50 000-odd k's of bashing around Australia you are going to break/damage/wear stuff out. I always tried to be careful with my kit—the budget was limited and flying in replacements was not usually an option. Nonetheless, I still notched up a disturbing tally.

- 2 crash helmets—bad luck as opposed to bad accidents (thankfully)
- 2 laptops—arguably motorcycles and laptops were never meant to mix
- 1 bike jacket—zip toggle came off and I ended up with a new jacket under warranty
- 1 DAT recorder—damaged by haemorrhaging batteries
- 1 DV camera—another victim of corrugations and vibration
- 2 top luggage box mounts—proving that after market products are always the weakest link
- 1 driving light—destroyed by bike falling over
- 1 headlight—destroyed by rocks from a passing truck
- 1 engine crash bar—cracked by who knows what
- 1 diff oil seal—caused by using the bike as a U-Boat
- 2 bike batteries—destroyed by the drain from the alarm
- 1 fuel line—destroyed by alarm siren chewing through the hose
- 10 tyres—destroyed by two laps of Australia
- 1 head torch—bad luck

L to R: The deaths of a luggage rack and a laptop.

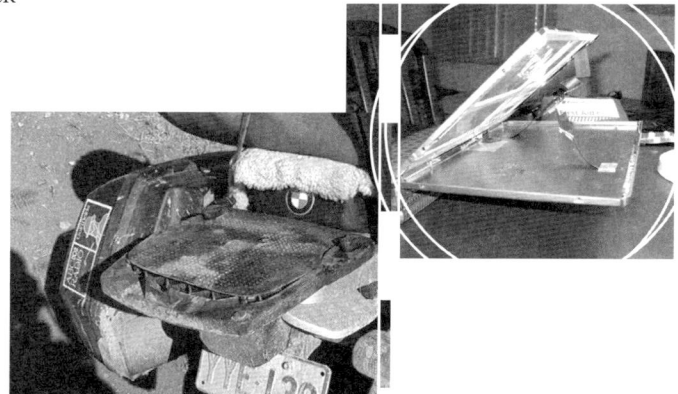

Trainspotting stats

LAP 2 OF AUSTRALIA—PART A

LOCATION	LITRES	ODOMETER READING	DISTANCE BETWEEN REFUELLING	
Gawler	21.53	1238		
Port Augusta	20.39	1554	316	
Woomera	21.23	1882	328	
Coober Pedy	27.00	2285	403	
Marla	21.70	2628	343	
Erlunda	17.13	2892	264	
Alice Springs	24.60	3254	362	New rear tyre
Kings Canyon	20.96	3603	349	
Uluru	27.63	4039	436	
Uluru	17.95	4201	162	Plus 10 litres in reserve tank
Warakurna	23.73	4569	368	
Tjukayirla	21.58	5084	515	
Laverton	20.42	5411	327	
Kalgoorlie	28.49	5830	419	
Menzies	8.10	5968	138	
Sandstone	18.27	6266	298	
Mount Magnet	12.65	6454	188	
Geraldton	25.31	6855	401	Oil change and new rear tyre
Mullewa	12.07	6924	69	
Murchison	13.26	7250	326	
Murchison	5.13	7328	78	
Mount Augustus	25.95	7754	426	
Ashburton Downs	9.00	8061	307	Bought petrol from station owner, short due to map error
Mount Tom Price	23.73	8223	162	
Mount Tom Price	18.66	8540	317	

LOCATION	LITRES	ODOMETER READING	DISTANCE BETWEEN REFUELLING	
Karratha	?	8925	385	? – Lost the flaming receipt!
South Hedland	22.91	9215	290	
Sandfire Flat	21.95	9543	328	
Broome	9.72	9955	412	
Broome	23.15	10026	71	
Derby	16.13	10284	258	
Mt Barnett	22.10	10653	369	
Home Valley	15.00	11076	423	
Kununurra	18.09	11216	140	Change oil because of submarine antics
Kununurra	19.74	11525	309	
Timber Creek	16.02	11757	232	
Top Springs	16.47	12021	264	
Larrimah	10.53	12361	340	
Katherine	?	12722	361	? – Damn Receipts!
Nauiyu/Daly River	10.18	12985	263	
Noomah	5.32	13198	213	
Darwin		13257	59	
				Fly & Freight back to Adelaide

TOTALS FOR JOURNEY 1

	713.78	12019		

LAP 2 OF AUSTRALIA—PART B

LOCATION	LITRES	ODOMETER READING	DISTANCE BETWEEN REFUELLING	
Adelaide	27.37	13800		Service and new tyres
Port Augusta	23.40	14150	350	
Leigh Creek	26.98	14529	379	
Maree	10.26	14697	168	
Mungerannie	18.92	14988	291	
Birdsville	26.20	15367	379	
Bedourie	13.53	15572	205	

LOCATION	LITRES	ODOMETER READING	DISTANCE BETWEEN REFUELLING	
Boulia	14.65	15777	205	
Dajarra	4.55	16054	277	
Mount Isa	24.41	16217	163	
Camooweal	11.02	16417	200	
Barkly	17.28	16698	281	
Heartbreak	23.60	17084	386	
Borroloola	14.48	17304	220	
Hells Gate	21.62	17634	330	
Burketown	16.43	17886	252	
Karumba	26.50	18283	397	
Burke & Wills RH	20.80	18579	296	
Julia Creek	18.13	18829	250	
Richmond	19.16	19104	275	New rear tyre
Winton	14.17	19345	241	
Longreach	14.54	19569	224	
Anakie	5.50	19963	394	
Emerald	23.08	20007	44	
Rockhampton	23.30	20318	311	Oil change
Maryborough	24.43	20724	406	
Gympie	8.79	20870	146	
Toowoomba	11.32	21241	371	
Brisbane	25.20	21441	200	
Goondiwindi	21.14	21801	360	
Moree	21.99	22109	308	
Tamworth	19.54	22401	292	
Newcastle	21.09	22724	323	
Lithgow	11.99	23121	397	
Orange	26.63	23334	213	
Parkes	8.96	23472	138	
Dubbo	12.17	23642	170	
Nyngan	11.70	23840	198	
Cobar	11.32	23983	143	
Wilcannia	18.18	24258	275	

LOCATION	LITRES	ODOMETER READING	DISTANCE BETWEEN REFUELLING	
Broken Hill	26.54	24629	371	
Oodlawirra	5.00	24951	322	
Hallett	13.20	25019	68	
Adelaide	27.56	25303	284	Service & new tyres

TOTALS FOR JOURNEY 2

	786.63	11503		

TOTALS FOR JOURNEYS 1 & 2

	1500.41	24 065 kilometres ridden		

LAP 1 OF AUSTRALIA

LOCATION	LITRES	ODOMETER READING	DISTANCE BETWEEN REFUELLING
Adelaide	14.96	1453	
Pirie	18.15	1738	285
Whyalla	16.83	2004	266
Lincoln	18.66	2310	306
Lincoln	11.33	2481	171
Lincoln	4.19	2520	39
Eliston	9.52	2745	225
Ceduna	17.68	3002	257
Yalata	15.83	3227	225
Nallarbor	9.30	3373	146
Mundrabilla	15.88	3654	281
Cocklebiddy	15.79	3910	256
Balladonia	13.22	4170	260
Norseman	12.39	4385	215
Kalgoorlie Boulder	15.41	4627	242
Norseman	11.52	4817	190
Esperance	14.53	5048	231
Esperance	12.13	5230	182

LOCATION	LITRES	ODOMETER READING	DISTANCE BETWEEN REFUELLING	
Ravensthorpe	14.49	5466	236	
Amelup	8.70	6220	754	
Albany	15.17	5824	-396	
Albany	18.67	6110	286	
Pemberton	17.83	6390	280	
Bunbury	18.97	6698	308	
Perth	19.96	6989	291	Oil change
Eneabba	17.00	7282	293	
Northhampton	12.51	7507	225	
Geraldton	13.27	7702	195	
Overlander RH	17.50	7995	293	
Shark Bay	12.85	8203	208	
Overlander RH	8.00	8339	136	
Carnarvon	11.79	8543	204	
Coral Bay	15.38	8800	257	
Nanutarra	16.30	9068	268	
Karratha	19.18	9363	295	
Port Hedland	20.21	9690	327	
Sandfire Flat	18.33	9991	301	
Roebuck Roadhouse	18.42	10291	300	New back tyre
Roebuck Roadhouse	11.95	10489	198	
Derby	12.70	10685	196	
Fitzroy Crossing	19.34	10986	301	
Halls Creek	20.47	11297	311	
Turkey Creek	9.86	11473	176	
Kununurra	12.85	11684	211	
Kununurra	21.88	12043	359	
Timber Creek	17.55	12293	250	
Katherine	18.28	12611	318	
Noonamah	10.02	12898	287	
Darwin	20.66	13071	173	Service
Darwin	19.23	13355	284	
Jabiru	17.26	13623	268	

LOCATION	LITRES	ODOMETER READING	DISTANCE BETWEEN REFUELLING	
Jabiru	6.73	13721	98	
Pine Creek	9.69	13982	261	
Katherine	17.23	14145	163	
Larrimah	12.82	14335	190	
Daly Waters	12.11	14443	108	
Cape Crawford	18.24	14718	275	
Barkly Homestead	22.90	15113	395	
Camooweal	14.49	15405	292	
Mount Isa	20.71	15706	301	
Julia Creek	15.88	15983	277	
Richmond	9.81	16142	159	
Hughenden	8.14	16263	121	
Charters Towers	15.89	16530	267	
Townsville	17.56	16801	271	
Carbutt	12.49	17022	221	
North Cairns	16.07	17266	244	
Daintree	10.00	17553	287	
Gordonvale	17.53	17742	189	
Wangan	10.00	18075	333	
Cairns South	21.51	18271	196	New front tyre
Ingham	14.22	18571	300	
Townsville	13.79	18717	146	
Prosperine	19.14	18992	275	
Mackay	9.32	19144	152	
Marlborough	16.03	19357	213	
Rockhampton	11.19	19553	196	
Miriam Vale	12.37	19764	211	
Bundaberg	13.61	19997	233	
Gympie	11.21	20182	185	
Ipswich	19.20	20480	298	
Dalby	16.22	20726	246	
Goondiwindi	14.93	20952	226	
Moree	9.38	21108	156	

LOCATION	LITRES	ODOMETER READING	DISTANCE BETWEEN REFUELLING	
Tamworth	21.91	21412	304	
Merriwa	12.57	21624	212	
Lithgow	15.69	21877	253	
Wyong	12.50	22127	250	
St Peters	18.09	22420	293	Service & new back tyre
St Peters	20.97	22647	227	
Marulan	14.47	22872	225	
Braddon – ACT	17.13	23099	227	
Tumut	10.10	23360	261	
Cooma	12.30	23456	96	
Jindabyne	10.64	23636	180	
Albury	19.95	23946	310	
Bright	16.10	24186	240	
Omeo	18.00	24443	257	
Woodside	21.70	24753	310	
South Melbourne	20.00	25071	318	
East Devonport	14.10	25246	175	
Launceston	20.47	25537	291	
St Helens	17.00	25797	260	
Triabunna	18.16	26066	269	
Sandy Bay	19.60	26301	235	
Perth	19.31	26588	287	
Burnie	13.25	26770	182	
Bendigo	21.81	27099	329	
Bendigo	14.04	27257	158	
Swan Hill	18.04	27541	284	
Mildura	21.01	27862	321	
Mildura	18.79	28159	297	
Renmark	20.69	28436	277	
Adelaide	23.14	28759	323	Service & new tyres
TOTALS	1766	27 306 kilometres ridden		

A day in the office

Travelling and working can be a bit of a juggling act, and I made it even more complicated with a tight schedule and the multiple mediums I was filing for. The following gives some idea of what an average day in the office involved. They didn't all go like this, but generally they were pretty mental, a lot of fun and I need a beer just thinking about it … So if you are interested in nuts and bolts kind of stuff, read on.

7.00   Throw something at the alarm clock—motorcycle boots tended to get the best result.

7.15   Shove some food down the throat & begin packing while chewing.

7.30   Pack for real.

8.15   Ablutions.

8.30   Finish packing—You might think I am exaggerating about the packing, but there was a rude amount of fragile technical gear that had to be carefully stowed in a minute space and in a manner that would ensure it would survive each day of bush bashing.

9.00   If I was lucky I might be fully packed by now. The last thing that remained was to put on the riding gear and make sure the helmet visor and sunnies were spotless—anal yes, but it helped because they would be caked in filth by the end of the day, not to mention it's always a bit of bonus being able to see where you're going.

9.15   Load the bike—clip on the panniers, tank bag, camping bag and top box. (The top box was my nemesis. After snapping it off twice, I resorted to lashing it on with extra straps to ensure it stayed put. I came to despise the straps as they chewed up time and made my recording equipment more difficult to access.)

9.30 Fuel up—I never filled up the night before as I'd previously lost precious fuel through the overflow pipe (this can happen if you are dopey enough to leave it on the sidestand). The morning fuel up ensured the tank was absolutely chokka—every millilitre counts with a motorcycle tank.

9.45 Leave (maybe)—The rule of thumb was that no matter how much I had nosed around the day & night before, the good stories only came to surface 5 minutes before I was due to leave. If this was the case, then everything ground to a halt. I would do a bit of qualifying and then track the person down, meet them and find out more about their story while trying to gain their confidence. Having their confidence was important—it was the difference between them talking naturally or not when the tape was rolling. Trying to cajole and engage with them wasn't always that easy when, in the back of my mind, I knew I had a few hundred k's of potentially slow going terrain ahead of me.

11.15 Role the Tape!—I wanted people to tell their stories. I was interested in recording conversations and stories, not flaming interviews. So the tape was left to roll for a bit so they would forget about what was going on, then slowly we'd ease into the chunky story stuff.

11.45 Wrap up the recording—I'd say thanks and apologise for having to bolt off in such a hurry. Usually they'd make me feel better by saying, 'You're going where? You should have left at 7 am to make that!'

11.50 Check rations. Make sure the water bag was topped up and buy three bags of lollies—it was the only food I could fit in the couple of empty crannies in my luggage.

12.00 Panic—the realisation that there was 300 to 400 k's in front of me and I should have been on the road ages ago.

12.05 Travel—Hopefully somewhere between 3rd and 6th gear … if in 6th it meant the going was well and I might make my destination before dusk; if in 3rd, I was in deep shit because the roos would be well out of bed before I made my destination, that was if I didn't run out of petrol because of the slow going. Once I had got over a bit of 'time management anguish' a big smile would creep across my face; I was doing something I loved and was passionate about—riding and travelling. To top it off, I was doing it for a job! A job that required me to meet all sorts of amazing people and soak up endless amazing countryside. I would pinch myself and then get on with the ride.

12.30 Photo stop—See something worthy of a photo, stop and take a few different shots. (Multiple shots were mandatory, as I'm average with a camera, as you may have worked out and so operate on the principle that at least one will turn out ok—the joy of digital photos!) Geography students were using some of the material I was gathering, so I would also enter the location of the image into the GPS (I didn't always remember to do this). Each day would be full of photo stops. In the two years on the road I took over 4000 images.

12.45 Back up to battle speed—If the road was semi decent I would probably be wrestling with the speed, trying to make sure I wasn't going too damn fast. That might sound dumb, but the flaming bike was over-engineered. It would flick easily over rough terrain and I wouldn't appreciate I was doing 130 until I looked at the speedo. And if it goes wrong with all that weight on board it goes very wrong. Once at cruising speed the ride would become a kind of meditation, I would be focused on the obstacles ahead while also trying to absorb the beauty of the country I was going through. Daydreaming wasn't a good plan, as I would miss spotting potential pain causers like bulldust holes. 100% concentration is a pretty taxing ask for the best Gen X-er and on top of this were the physical demands. Much of the riding would be done standing up; the footpegs are the safest place to be as it lets you use your weight to efficiently manoeuvre the bike. To complete the mix, it was usually nice and toasty with the temperature in the mid 30s. The irony about all this was the harder the ride the more rewarding the day's travelling—masochism anyone?

3.00 Lolly break and maybe a bit of 'venting'—Bladder functions though were a rarity as most of the fluid had been sweated out. Usually this break in the day was tied in with a photo stop. The breaks were short as I would generally be running late. An added incentive to get going was the heat slowly broiling me inside the riding gear. Probably the real reason the breaks were short was because I was always itching to find out what was around the next corner.

3.10 More tape rolling. Get going and muse that it's usually around this time (i.e. still a long way to go and with little daylight remaining) that I bump into someone that is really interesting and would make a great story. I stopped and chatted to most people I came across when I was in remote areas: it was the only way to find stuff out and source stories.

3.30 Shadow Watch—The dry season/winter is really the only time to head north. The only downside is that there is less daylight. On my schedule this meant I was always riding in late afternoon light, when the shadows from the trees were long and thick across the track. The shadows usually hide big potholes in monsoon country and the ailing light also meant skippy and his mates were starting to wake up.

4.30 Hopefully within 80 k's of the final destination—I would begin conducting endless repetitive mental calculations on:
• amount of daylight left versus distance remaining
• distance remaining versus fuel remaining.
The mental gymnastics to reconcile the numbers usually paled into insignificance as a stunning sunset crept across the road.

5.30–6.00 Roll into wherever is home for the night—If I had the energy I would castigate myself for cutting it so fine and not leaving any room for stuff-ups like flat tyres. (Admittedly in two years I didn't get a flat, but it still wasn't an excuse for not leaving a margin for error.) Usually at this time of day I would also think 'If Mission Central really knew what was going on out here, there is no way they would ever let me out of the office again.' After the self-berating I would begin beaming about having had another wicked day on the road.

6.00–7.00 Unload, climb out of riding clothes, scam a shower to wash off the sweat, grime and bull dust. Unpack laptop, cameras, etc. and set up the production suite. The final bit of 'suite assembly' involved changing the crummy 40-watt light bulb that is always used in dongers for one of the 100-watt globes I carried; Call me petty but I liked to be able to see the laptop without going blind.

7.00 Scrounge up something to eat—usually an entertaining experience for a vego travelling the outback.

7.30–8.00 Begin churning out the stories. From this time, to when I'd collapse in the cot, a mix of the following would happen.

Data Transfer
- upload images from digital still camera into laptop
- upload days readings from GPS into laptop
- upload any audio/visual recordings from the day into the laptop.

Website Images
- Select best images for the web and crunch down to a web friendly size. Write appropriate text and attach the GPS coordinates I had remembered to record. Split up data and prepare for file transfer back to ABC Mission Central.

Edit Recordings
- TV—fortunately there were only a few of these. I say this because digitising and editing one of these suckers can be a bit of a mission.
- RADIO—Edit up the story (with my long recordings this was often an epic task), record my voice links, mix in the music and sound effects, crunch it into a high quality MP3 file for transmission back to base, write an introductory script for announcers.

Write Up School Visits
- If I visited a school during the day an entry for the web was required. The idea of these was to give an insight into school life in this corner of the country.

Charge All Batteries
- Mobile & Sat phone/Video Camera/Laptop, etc.—Most remote locations use a generator to produce their power. I would spend the night praying that the tired old diesel plant, which was making the lights flicker and dim, was not dodgy enough to fry all my equipment.

Bean Counting
- Enter in all fuel purchase details and mileage readings for publishing on the website
- Spreadsheet all expenses incurred throughout the day.

Data Management
- Backup all data
- If I was in a place that had a phone line I would transmit the data back to base.

12.30–1.00 AM Hit the pillow exhausted, but wondering who I was going to meet and what I was going to see the next day. I had the best job in the country.

L to R: One pristine set of wheels, seconds before it started bashing its way around the country.

One of the more useful directions in the deserts north of Birdsville.

More trainspotting stuff

Duration of project (in years)	2
Weeks on the road	43
Kilometres travelled	51 371
Number of TV items	16
Number of radio packages	62
Number of live radio updates	139
Words written for web	104 629
Photos used on web	1516
Litres of fuel used	3266
Number of fuel stops	198
Radio documentary series	1
Books	1
Number of people interviewed	200+